T0233949

Lecture Notes in Computer Science   11667

Commenced Publication in 1973
Founding and Former Series Editors:
Gerhard Goos, Juris Hartmanis, and Jan van Leeuwen

FoLLI Publications on Logic, Language and Information

Subline of Lectures Notes in Computer Science

More information about this series at http://www.springer.com/series/7407

Jennifer Sikos · Eric Pacuit (Eds.)

At the Intersection of Language, Logic, and Information

ESSLLI 2018 Student Session
Sofia, Bulgaria, August 6–17, 2018
Selected Papers

 Springer

Editors
Jennifer Sikos
Universität Stuttgart
Stuttgart, Germany

Eric Pacuit
Department of Philosophy
University of Maryland
College Park, MD, USA

ISSN 0302-9743   ISSN 1611-3349   (electronic)
Lecture Notes in Computer Science
ISBN 978-3-662-59619-7   ISBN 978-3-662-59620-3   (eBook)
https://doi.org/10.1007/978-3-662-59620-3

LNCS Sublibrary: SL1 – Theoretical Computer Science and General Issues

This Springer imprint is published by the registered company Springer-Verlag GmbH, DE
part of Springer Nature
The registered company address is: Heidelberger Platz 3, 14197 Berlin, Germany

Preface

It is our pleasure to present this volume of selected papers from the Student Session, held at the European Summer School in Logic, Language and Information (ESSLLI). The publication of the ESSLLI Student Sessions' proceedings was previously a biannual tradition, but this year we have collated selected papers from a single Student Session to give students the opportunity to accelerate the public release of their work. The 14 papers presented in this volume were selected among 24 papers presented by talks or posters during the Student Session at the 30th edition of ESSLLI, held at Sofia University "St. Kl. Ohridski" in Sofia, Bulgaria. The papers are extended and revised versions of the papers presented at ESSLLI, and have all been subjected to a second round of blind peer review. The papers cover vastly different topics, but each fall in the intersection of the three primary topics of ESSLLI: logic, language and computation.

The Student Session is a collaborative effort between a group of student co-chairs, the Local Organizing Committee, FoLLI's ESSLLI Standing Committee, area experts who help to guide student co-chairs through the chairing and reviewing process, and not least the reviewers who provide students with valuable feedback. Student authors are thus given a venue to explore novel ideas and receive input from a community of experts and peers. This year, students had the opportunity to experience Bulgarian culture and history through various excursions organized by the local Organizing Committee, including tours of the historic town of Plovdiv and the largest Eastern Orthodox monastery in Bulgaria, the Rila Monastery.

Apart from continuing the tradition of providing a forum where young researchers may present their work in a constructive and supportive environment, we were also very happy to be able to continue the tradition of concluding the Student Sessions with prizes for the Best Paper and Best Poster. The determination of the winners for these prizes were based on a combination of their paper's reviewer scores and judgments by the Area Experts, where the students with the highest combined scores were this year's recipients. These prizes were awarded to two of the student authors present in this volume—Lauren Edlin received the Best Poster award for her paper "Simulating the No Alternatives Argument in a Social Setting," and Zhuoye Zhao received Best Paper award for his paper "Interpreting Intensifiers for Relative Adjectives: Comparing Models and Theories." Best Paper/Poster prizes were generously supplied by Springer, and we would like to extend our gratitude to Springer for supporting the Student Session consistently for so many years. We sincerely hope you find this volume of papers as exciting, novel, and innovative as we have, and above all we thank the student authors for their hard work and their dedication to producing research of such high quality.

May 2019

Jennifer Sikos
Eric Pacuit

Organization

Editors

Jennifer Sikos   University of Stuttgart, Germany
Eric Pacuit   University of Maryland

Program Committee Chair

Laura Kallmeyer   Düsseldorf University, Germany

Local Organizing Committee

Petya Osenova   Sofia University, Bulgaria
Kiril Simov   IICT-BAS, Sofia, Bulgaria

Folli Steering Committee

Valentin Goranko   Stockholm University, Sweden
Michael Moortgat   Utrecht University, The Netherlands

Program Committee

Robert van Rooij   University of Amsterdam, The Netherlands
Natasha Alechina   University of Nottingham, UK
Michael Moortgat   Utrecht University, The Netherlands
Peter Sutton   Heinrich Heine University Düsseldorf, Germany
Mira Grubic   University of Potsdam, Germany
Melissa Fusco   Columbia, USA
Thomas Weskott   Georg-August-Universität Göttingen, Germany
Ivano Ciardelli   Ludwig-Maximilians-Universität München, Germany
Sarah Zobel   University of Tübingen, Germany
Jacopo Romoli   Ulster University in Belfast, UK
Daniel Altshuler   Hampshire College, USA
Hedde Zeijlstra   Georg-August-Universität Göttingen, Germany
Fernando R.   University of Amsterdam, The Netherlands
    Velazquez-Quesada

Contents

Simulating the No Alternatives Argument in a Social Setting

Lauren Edlin[(✉)]

Faculty of Sociology, Bielefeld University, Bielefeld, Germany
`ledlin@uni-bielefeld.de`

Abstract. This paper offers an initial investigation into how the number of choices available to individual agents may influence choice at the group level by formalizing and simulating a social version of the No Alternatives Argument (NAA). The Social NAA assumption predicts that strength of belief in the most strongly held hypothesis in a group of agents will increase when the number of available hypotheses decreases. Social network simulations using connected Bayesian networks show that this assumption can be violated, but infrequently. Implications of the Social NAA assumption and when it holds in social networks are discussed, and future work is outlined.

Keywords: No Alternatives Argument · Bayesian epistemology · Social epistemology · Social networks · Opinion dynamics

1 Introduction

This paper is a preliminary investigation into how the number of choices available to individual agents in a group setting may influence choice at the group level. Empirical studies from organizational and consumer psychology (e.g., [20,23]) provide evidence that the number of choices is indeed an influential factor on individual agent's decision-making. The relation between the number of choices for the individual and the outcome at the *group* level, however, is not considered in most formal models of social decision-making, most likely because this relation is not straightforward – a descriptive account is difficult to generalize because decision-making processes depend on the context and type of decision being made. Nevertheless, there are many contexts in which understanding more about the relationship between group decision-making and the number of choices available to individuals in that group could be useful, such as understanding voting dynamics in democratic elections, interpreting a hypothesis choice in a scientific community, or understanding consumers' buying behavior.

The research described here therefore models, tests, and discusses one assumed descriptive relation between the number of choices *regarded as feasible to choose* by individuals and their group-level choice, in order to explore how this relation can be formally approached in future work. The tested assumption

© Springer-Verlag GmbH Germany, part of Springer Nature 2019
J. Sikos and E. Pacuit (Eds.): ESSLLI 2018, LNCS 11667, pp. 1–20, 2019.
https://doi.org/10.1007/978-3-662-59620-3_1

is: as the number of choices regarded as 'feasible' by individual agents decreases, the strength of belief in the group level choice will increase, as indicated by the number of agents having chosen it. This assumption is adapted from Dawid, Hartmann and Sprenger's (DHS) [7] formal proof for the validity of the No Alternative Argument (NAA) in the framework of Bayesian epistemology (e.g., [6]). The NAA posits that the absence of found or developed alternative choices evidences that the one available choice is best, correct or true. This argument is used in a variety of contexts, with a notable political example being the former UK Prime Minister Margaret Thatcher's slogan for her economics policies being "There is no alternative", often shortened to 'TINA' [4]. In order to prove that this argumentation strategy is rational, DHS construct a probabilistic analysis that heavily relies on the intuition that the fewer the number of available choices there could be, the strength of belief in the one choice (that is, the most strongly believed choice by default) should increase, or at least not decrease. A new model is developed here in the same framework employed by DHS to define and test the reformulation of this assumption for an explicitly social setting (given above). From here on, the tested assumption will be referred to as the 'Social NAA assumption'.

The structure of the rest of the paper is as follows: Sect. 2 provides more precise conceptualizations of 'feasible' choice and group choice, and discusses examples of such choices that inform the model developed in later sections. Section 3 describes DHS's formal argument for the NAA in detail, and provides critiques of their paper that also informs the model developed here. Section 4 describes the new Bayesian network model and offers a formal conception of the Social NAA assumption. Section 5 outlines the parameters, including the number and configurations of agents in a social network, and the results of several simulations over various versions of the model. Section 6 concludes with a discussion of the results, the current shortcomings of the work, and how to address them in future work.

2   Most Feasible Choice: From the Individual to the Group

2.1   Individual Ranking of Feasible Choices

When an individual is prompted to make a decision, he or she will make whichever choice is perceived as most feasible for the decision at hand, often after limiting the decision to a subset of choices determined as feasible to choose [3,20]. A *feasible choice* is taken to mean a choice that the agent believes is possible to make, given the available information and circumstances. A choice may be more or less, and indeed most or least feasible—in the probabilistic model developed in later sections, feasibility is described as a real-valued number, or, in other words, a subjective judgment of a choice's probability.

The conception of 'most feasible' choice is similar to inference to the best explanation (IBE) (e.g., [12,15]). IBE (or abduction) is reasoning based on the 'explanatory fitness' of a choice, such that one is rational to decide on the choice that best explains the available evidence or circumstances. A scientist, for instance, will support a hypothesis that gives the best explanation for the data at hand – while another hypothesis may offer a better account for predicting future observations, the choice that provides the most coherent explanatory understanding of the data is argued to be most likely the true, and therefore the most rational hypothesis to support.

While IBE is ubiquitous in everyday life, and coincides with the above conception of feasibility (an individual may deem a choice most feasible because it has the greatest explanatory fitness), this research conceives of choice rank and decision-making only through feasibility rather than IBE for two reasons. First and fundamentally, feasibility is thoroughly descriptive, while IBE is prescriptive. The purpose of this research is to model individuals' decision-making in a group setting, rather than argue for how individuals *should* reason about which choice is the closest proximity to truth. Therefore, the model developed in later sections does not stipulate an objective measure on which choice is meant to be true or best. Instead, the probabilities attached to each choice are interpreted to be subjectively assigned belief states based on factors including but not limited to explanatory fitness – other relevant factors such as psychological biases or, more relevantly, the number of available (but perhaps less feasible) choices, may also influence decision-making.

Secondly, because ranking choices by feasibility is purely descriptive, it avoids the problem of there not being a 'good enough' choice to rationally make. Proponents of IBE emphasize that even the strongest explanatory choice may not be 'good enough' to rationally choose (e.g., [15] – p. 154)[1] – that is, the choice with the highest explanatory fitness may not pass a threshold of plausibility (e.g., it is not more likely to be true than not), and under IBE deciding on this choice is irrational. Under feasibility, however, it is possible to decide on such a choice, or even decide to not make a choice at all. Again, there simply isn't an objective best choice that should be chosen.

In sum, feasibility as described here is broadly decision-making informed by the numerous factors that influence preference rankings for a set of choices. It can be plausibly argued that this conception is *too* general to be useful: scientists deciding between hypotheses is very different to an electorate casting votes to decide the next national leader or ruling party, and these differences will most likely influence the relation between number of choices and group-level choice. Nevertheless, a simple conception of ranking that fits with probability assignments as a subjective measure is beneficial for the early stages of this research – the hope is that this general formulation may reveal promising avenues of approaching this problem in more specific contexts.

---

[1] Thank you to an anonymous reviewer for bringing this point to attention.

2.2 Changes in the Number of Choices and Group-Level Choice

What is meant by *changes in the number of choices* and how choice is understood at the group level needs to be elaborated, as the purpose of this research is to explore how changes in the number of choices perceived feasible by individuals influences choice at the group level.

As previously described, individuals will rank the choices from most to least feasible when prompted to make a choice. Some choices will not reach the (subjective) threshold of feasibility, and are essentially dropped from the original set of choices. Evidence for such 'pre-screening' of choices is supported by psychological literature (see [3, 20]). This is the primary way in which the number of perceived feasible choices may change: through an individual and subjective sanctioning of the number of available choices. Since choices are usually dropped during this screening process, the number of perceived feasible choices often decreases. Choices may be added, however, as more information about the choices is received by an individual and the choice set is re-assessed. This implies that choice screening is an iterative process, and new rankings are constantly being established as new information is presented.

Additionally, in some cases choices are explicitly added or detracted by external circumstances. This type of change in choice number also influences feasibility rankings. Take the example of the U.S. primary and general elections. In the primary election, voters choose from a set of potential candidates representing their supported party.[2] A voter will rank the candidates based on information from a variety of sources and voting strategies, keeping in mind that the chosen candidate will be up against a new opponent in the general election. Due to the impending general election that will fix the number of choices to two candidates, it is possible that a candidate who would normally be considered unfeasible for various reasons (such as their less desirable policies, or being out of kilter with their party's traditional messages) becomes the most feasible option to a particular voter because of the potential opponent candidate, thus prompting the voter to decide against their previous candidate ranking. After the primary, the number of choices the voter must consider in the general election becomes two, and again the voter re-calculates their candidate choice.

The election example also presents a simple measure of group choice by majority vote. While group choice can be measured in a number of ways (see [16] for an overview), a majority vote is considered adequate for the purposes of this research, which is interested in descriptive decisions between choices rather than explore what parameters of group choice may lead to the best, true, or most

---

[2] Who is permitted to vote in a primary election depends on the state. Open primaries, for instance, allow for any registered voter to vote in any party, regardless of party affiliation. Closed primaries, on the other hand, only allow for registered members of that particular party to vote. The type of primary is not necessary to specify for this example, although it does effect the voters' calculation of which is the most feasible choice – a voter may, for example, employ a strategy of 'crossing over' to vote in the primary for the party he or she is not affiliated with to try to choose the opponent they would prefer to face in the general election.

useful choice. Majority vote is also considered valid for non-political contexts too, for example consumers 'voting' for their favourite brands by buying them, and in cases where group choice may not be as explicit such as when a scientific community expresses preferences among various hypotheses.

Before turning to a description of the Social NAA assumption in light of these ideas, it is important to point out that an essential aspect of choice feasibility ranking, as implied by the election example, is that it is a social process. Individuals determine their feasibility ranking using information from two broad sources: evidence from their own experiences in the world, and information from social interactions and opinion sharing with others in their social network. Per the election example, individuals are informed about the candidates through sharing opinions, consuming media, and weighing up this information with their beliefs and values that have been shaped through experiences over time. Therefore, feasibility ranking and the number of choices deemed feasible depend to some extent on social interactions, and the model developed in later sections must account for opinion sharing dynamics in order to adequately test the Social NAA assumption.

This formalization, which in turn informs the model developed here to test the Social NAA assumption, is discussed in the next section. It is important to note that DHS constrain their work to a scientific context, and therefore refer to choices as 'hypotheses' for a given set of empirical data.

3   Dawid, Hartmann and Sprenger's NAA

Dawid, Hartmann and Sprenger (DHS) [7] consider the case in which only one hypothesis is found that accounts for a set of evidence, despite extensive efforts to find more, and formalize the relationship between the strength of belief in one hypothesis and the number of alternatives available. In the context of scientific inquiry, a situation of no alternatives may occur when there is little empirical evidence available to construct more than one hypothesis, as is common in fields such as paleontology and theoretical physics. DHS aim to show that it is rational (or at least not irrational) for a scientist to strengthen their belief in the one available hypothesis *because* it is the only found hypothesis[3] – however, observations about the number of hypotheses available to choose from do not typically constitute empirical evidence for or against scientific hypotheses. DHS therefore must bring in other relevant factors to allow this observation to serve as indirect or 'non-empirical' evidence.

The primary factor that provides this indirect link relies on the number of potential hypotheses 'out there' that could account for the data. DHS's motivating intuition is that the larger the number of alternative hypotheses, the less likely it is that the hypothesis found by scientists is the correct one. This is driven by the belief that if there are actually a large number of alternative hypotheses,

---

[3] Although the owner of this belief state, whether it is that of the scientific community as a whole or the individual scientists, is not clarified by DHS. This point will addressed in the new model developed in later sections.

then it is plausible to expect that scientists would have found more than one after extensive efforts to do so. The Social NAA assumption proposed here has a similar underlying motivation – both claim a decrease in the number of choices (or hypotheses) increases the strength in belief in a particular choice.

In addition, DHS point to other factors that affect whether more than one hypothesis has been found, which includes the technological and computational resources available to the scientists, the cleverness of the scientists, and the difficulty of the scientific problem being investigated. These factors coincide with the empirical and social factors that influence an individual's choice feasibility ranking, as discussed in the previous section. However, while DHS claim that these cumulative factors influence whether more than one hypothesis is found, they also stipulate that these factors have an influencing effect that is independent from the number of potential hypotheses – in other words, the number of hypotheses 'out there to be found' is not determined by or beholden to the social aspects of the scientific communities that develop and apply them. Instead, the influence of the number of hypotheses and the cumulative factors of the scientific community are described as separately influencing whether more than one hypothesis has been found, as seen in the formalized proof below.

The formalized proof for the NAA in the context of Bayesian epistemology proceeds as follows.[4] Let (Ω, F, P) describe a probability space with sample space Ω as a set of possible worlds, F as a σ-algebra delineating subsets of Ω, and a countably additive probability measure P on F, such that the following are P-measurable:

The binary propositional variables

T: The hypothesis H is correct for the set of data at hand[5]
$\neg T$: The hypothesis H is not correct for the set of data at hand
F_A: The scientific community has found not yet found an alternative to H
$\neg F_A$: The scientific community has found a viable alternative to H

And the functions

$Y : \Omega \rightarrow N$, where $Y(\omega)$ is the number of alternatives to H (in world ω)
$D : \Omega \rightarrow N$, where $D(\omega)$ is the degree of difficulty of the scientific problem addressed by H
(where N denotes the set of non-negative integers)

Y stipulates the number of alternative hypotheses that can account for a given data set. D describes the cumulative factors of difficulty, such as technological resources and social dynamics of the scientific community, that influence whether scientists have found more than one hypothesis.

---

[4] This iteration of the formal proof borrows from [13].

[5] In the original paper, the proposition T is "the hypothesis is empirically adequate". DHS set up their proof to make inferences about the empirical adequacy of the found hypothesis rather than its truth to avoid the possibility of constructing infinite arguments with superfluous additions, which they claim threatens the validity of the NAA. This is not a point of contention for the new model developed in later sections because the concern is decision-making rather than inference.

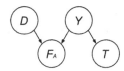

Fig. 1. Bayesian network representation of the NAA

DHS determine under which conditions F_A confirms hypothesis H, or as typically described through Bayesian confirmation, when $P(T \mid F_A) > P(T)$. In other words, DHS must show that there is greater credence that H is correct given the fact that scientists have not found an alternative to H, than to H being correct without the additional observation of it being the only possible hypothesis. Since T and F_A are not empirically or probabilistically related to each other, the mediating factors of Y and D are required. To prove $P(T \mid F_A) > P(T)$, five assumptions are described that give structure to a Bayesian network by assigning prior and conditional probabilities:

A1: T is conditionally independent of F_A given Y; i.e. $P(T \cap F_A \mid \{Y = n\}) = P(T \mid \{Y = n\})P(F_A \mid \{Y = n\})$ for every $n \in N$.

A2: Y and D are independent.

A3: $P(F_A \mid \{Y = k\} \cap \{D = j\})$ is non-increasing in k for every fixed j as well as non-decreasing in j for every fixed k.

A4: $P(T \mid \{Y = k\})$ is non-increasing in k.

A5: There exists at least one pair $i, j \in N$ such that
  1. $j > i$
  2. $P\{Y = i\}, P\{Y = j\} > 0$,
  3. $P(F_A \mid \{Y = i\} \cap \{D = k\}) > P(F_A \mid \{Y = j\} \cap \{D = k\})$ for some $k \in N$,
  4. $P(T \mid \{Y = i\}) > P(T \mid \{Y = j\})$.

A formal proof is provided in the appendix of DHS's paper [7] that shows if A1–A5 hold and Y takes values in non-negative natural numbers, then $P(T \mid F_A) > P(T)$.

Assumptions A1 and A2 give rise to the Bayesian network in Fig. 1, which depicts the indirect relation between T and F_A through Y. A2 stipulates that the number of hypotheses is not determined by the cumulative factors described in D. A3 and A4 are of particular interest because they present a formal interpretation of the motivating intuition behind DHS's proof, that the larger the number of alternative hypotheses the less likely it is that the hypothesis found is the correct one. A3 gives the conditional influences of D and Y on F_A: an increase in the number of potential hypotheses does not decrease the possibility of finding an alternative to H, and an increase in the difficulty of the scientific problem does not increase the likelihood of finding an alternative to H. A4 gives a strong interpretation of the intuition that an increasing number of potential hypotheses does not mean it is more likely that scientists have found the correct hypothesis, or, in other words, as the number of alternatives decreases, the likelihood of H being the correct one increases. Finally, A5 states that there is at least one case in which a large number of potential hypotheses makes it less likely that H is correct.

3.1   Criticisms of the NAA: Informing a Social NAA Assumption and Model

DHS's NAA is used as a starting point for developing a new Bayesian network model to test the Social NAA assumption. Recall that the primary interest in the NAA in this research is its motivating intuition that the fewer the number of potential hypotheses, the more likely it is that the one hypothesis can be inferred to be correct. Explicitly including the number of potential hypotheses as a variable, and its influence on other belief states in an epistemic model, is a novel and interesting contribution. Critiques of DHS's work, however, show where the assumptions in their model fall short. The critiques of the NAA that are conducive to shaping the model to test the Social NAA assumption are briefly given here.

Firstly, the assumption that the social processes inherent to scientific practices do not play a role in hypothesis formation and choice is highly contentious. Herzberg [13] points out that while there may be an interesting class of cases where Y and D are mostly independent, assumption A2 is overall difficult to motivate, especially when hypothesis legitimacy and choice is known to be a socially complex process (e.g., [14, 18]). Herzberg offers a generalized NAA proof that is capable of proving $P(T \mid F_A) > P(T)$ without A2.

Secondly, it is unclear whose belief states are described by the probabilities in DHS's Bayesian network. The Social NAA we develop here will specify individual actors to allow for opinion sharing and information exchange to address the issue of *whose* beliefs are under discussion, those of the individuals or the community as a whole. As will be shown below, the specification of individual agents in a network makes it possible to determine group choice through a simple majority vote, whilst also describing the choice of the individuals in that network.

Finally, both subjective and objective interpretations of the probabilities in DHS's proof have their problems with regards to the validity of the NAA. Van Basshuysen [2] plausibly argues that an objective interpretation leads to the NAA's invalidity because it presupposes non-empirical evidence—this is problematic as the NAA proof is meant to show the validity of the non-empirical observation that only one hypothesis has been found. Van Basshuysen claims there is no reason to assume a relationship between the number of hypotheses and scientists increasing their belief in the one hypothesis they found, besides the fact that the scientists are experts and their opinions about hypothesis choice merit recognition. However, this reason is an appeal to authority, which is a non-empirical influence on choice. The appeal to non-empirical evidence ultimately stems from the number of alternative hypotheses under an objective interpretation being understood as fixed entities 'out there' to either be uncovered or found, or remain out of reach of comprehension. This conception of hypothesis formation and choice is overly simplistic, and prevents more reasonable accounts of hypothesis formation and choice that involve social processes (as argued for in previous paragraphs). For this reason, it is more plausible to consider the number of hypotheses as subjective judgments made by the scientific community.

Turning to a subjective interpretation, Van Basshuysen points out that this is also undesirable for the NAA because it becomes a descriptive account of how some scientists may reason when there is only one hypothesis available. This is undesirable for DHS, as their goal is to prove the validity of the NAA. A subjective interpretation, however, is desirable for a model to test the Social NAA assumption. This is because, as described in Sect. 2, the Social NAA assumption does not test an inference relation, but rather a decision made from a set of choices. In other words, the NAA attempts to show inference to a *true* hypothesis, while the Social NAA model we describe here does not specify a true or correct choice and instead looks to describe a decision-making process.

The next sections clarify how the NAA can be scaled to a social setting through the Social NAA assumption, define a formal model which can test it and describe a set of computational simulations run to determine if the assumption holds across a number of cases. Taking the above criticisms into account, the model will incorporate opinion and information sharing between individuals in a social community. Although the Social NAA assumption tests a general decision on a set of choices across contexts, the term 'hypothesis' in maintained in continuity with the NAA, but is taken to stand for 'feasible choice' across contexts beyond scientific practice.

4   Social NAA: Agents in a Social Setting

For a given community making choices where choice preference may be communicated between individual agents, we are interested in exploring the following assumption in Definition 1.

Definition 1. *The Social No Alternatives Argument assumption: As the total number of choices regarded as 'feasible' by agents in a community decreases, the strength of belief in the group-level choice will increase, or at least not decrease, as indicated by the number of agents having chosen it.*

We chose this particular assumption for testing because it represents a core assumption of the original NAA, and describes the empirically supported act of 'pre-screening' choices. That is, as individuals drop choices that do not reach their threshold of feasibility, it is plausible to expect that the majority vote will increase as the set of choices to decide from inherently becomes smaller. Although this assumption may at first appear trivial or self-fulfilling, it is not guaranteed to hold for any reduction in choices other than from two to one, where this is a trivial case of the NAA in our setting. We are focusing on whether the function from the number of alternatives to the strength of belief in the choice is non-decreasing, which is required to prove the NAA (A4 in DHS's NAA proof) – here we aim to test the core assumption's validity in a social setting, rather than stipulate it.

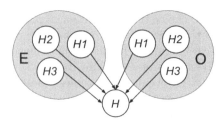

Fig. 2. A Bayesian network representing the belief states of one individual agent considering three hypotheses.

We explore this through simulating a community network consisting of multiple Bayesian networks which represent individual agents able to communicate with one another—they are connected in a way to represent opinion sharing between specified agents. The Social NAA assumption is tested dynamically over time in simulations. As the individual agents update their strength of belief in hypotheses through opinion sharing over time, some hypotheses drop out of individuals' feasibility thresholds by dropping to zero probability. After the agents share their opinions with an iteration of inference, a majority vote is calculated to determine whether the Social NAA assumption was upheld at that time-step in the simulation.

The model is described below, beginning with defining individual agents before scaling up to the community network and dynamics over time.

4.1   Individual Agent Networks

We define an individual agent as a Bayesian network as in Fig. 2. The categorical conditional variable H represents a probability distribution over the set of hypotheses under consideration by the agent. If three hypotheses are under consideration, for instance, then the values of H are the set $\{H1, H2, H3\}$. The probabilities assigned to the values constitute a standard probability distribution and therefore sum to 1.

Two types of priors determine the distribution in H, which are represented by E variables and O variables. E variables (with 'E' standing for "experience") represent the credence afforded to each hypothesis according an agent's own experiences, experimentation, values and biases up to a specific point in time. O variables (with 'O' standing for "others") represent the credence afforded to each hypothesis according to the collated opinions of other agents in the network which can influence that agent.

The number of E and O variables reflects the number of hypotheses under consideration. Keeping with the example of three hypotheses in H, the network would include three E variables and three O variables, which will be respectively labeled E_{H1}, E_{H2}, and E_{H3} for the E variables, and O_{H1}, O_{H2} and O_{H3} for the O variables.

Each E and O variable is a categorical, binary variable. For each variable, the value T (equivalent to the propositions E_{Hi} or O_{Hi} for a given hypothesis H_i) means the hypothesis indicated is perceived to be a feasible option. The value F (equivalent to the propositions $\neg E_{Hi}$ and $\neg O_{Hi}$ for a given hypothesis H_i) means the hypothesis indicated by that variable is considered an unfeasible option for whatever reason (e.g., it is not supported with sufficient evidence, it is not well known to the agent, there is believed to be a low chance that this option will have an impact, etc.). For each variable, for a given hypotheses H_i it will always be the case such that $p(E_{Hi} = T) + p(E_{Hi} = F) = 1$ and $p(O_{Hi} = T) + p(O_{Hi} = F) = 1$ in line with classical probability theory.

The E and O variables are priors that are used to compute the posterior probabilities in H as will be described below.

4.2 Conditional Probability Assignments Weighting Self and Other Beliefs

Agents take into account the opinions of others and bias their own preconceived beliefs to various extents. Some agents, for instance, may readily take on board the opinions of others in cases where they believe themselves to be less informed or consider others more expert, while other agents may consistently disregard the opinions of others and stick to their own experiences to support their beliefs. Such differences in bias between agents are simulated through different weights given to the E and O variables when calculating the conditional probabilities of H.

This paper describes different types of agent based on their weights as either *balanced*, *mule*, or *sheep* type agents. A balanced agent gives equal credibility to its own preconceived opinions and the opinions of others, and therefore each E variable has a weight of 0.5 applied to it in H, and each O variable also has a weight of 0.5 applied to it in H. A mule agent gives greater credibility to its own preconceived opinions, and therefore gives each E variable a weight of over 0.5 in H and each O variable a weight of under 0.5 in H. A mule type agent is consequently less likely to change its preferred hypothesis in light of its neighbor's opinions. Finally, a sheep agent is weighted in the converse way to the mule, with a weight of under 0.5 given to each E variable in H and a weight of over 0.5 given to each O variable in H. A sheep type agent is therefore more likely to change its preferred hypothesis in light of the opinions of its neighbors and is a follower. While we talk about these three types in the discussion below, in practice we do not stipulate each agent to be of these types in the experiments, but use a random weight $W \in [0, 1]$ such that each agent gives W weight to its O variable hypotheses and weight $1 - W$ to its E variable hypotheses. This random assignment results in agents with a real-valued range across the three types.

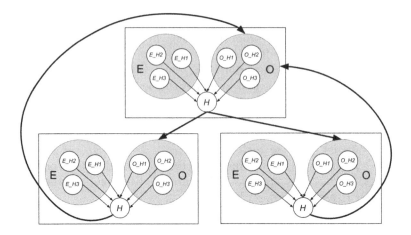

Fig. 3. A three-agent network, with each agent considering three hypotheses.

4.3 Scaling up to a Social Network: Collating O Variable Values from an Agent's Influencing Neighbors

The probabilities of the O variables for agents in a community network are determined by the values in H from specified neighbor agents, as we assume some function of H is an output variable that can be communicated by agents, while E and O variables are part of an agent's private reasoning.

To define the notion of neighborhood, we characterize the network as a directed graph. A *parent* agent refers to an agent that influences another's O variables, while a *child* agent is influenced by a parent agent. Each agent aggregates the top H values of all its parents, and uses that to set new values for its own O variables. For example, if an agent has three parents where two of those parents preferred $H1$ and one preferred $H2$, the child agent would set its O_{H1} variable's distribution as $p(O_{H1} = T) = \frac{2}{3}$ and $p(O_{H1} = F) = \frac{1}{3}$, and set its O_{H2} variable's values to $p(O_{H2} = T) = \frac{1}{3}$ and $p(O_{H2} = F) = \frac{2}{3}$. All the other O_{Hi} variables would have the distribution $p(O_{Hi} = T) = 0$ and $p(O_{Hi} = F) = 1$ (i.e. the other hypotheses have 0 probability according to the agent's parents). Given the O variables rely on the votes of its parents, it is easy to see how hypotheses can drop out after initial consideration, particularly for sheep agents where more weight W is given to the O variables than not, as there is no probability mass reserved for hypotheses in O with no support.

Figure 3 depicts a three-agent network, in which the top agent is the parent to the two lower child agents, influencing their belief through their O variables. As the recursive loop from the children's H to their parents shows, the child agents also inform the original parent agent's O variables. However, since Bayesian networks are by definition Direct Acyclic Graphs that forbid looping configurations, the influence of the two lower agents on the top agent must be done in a a subsequent belief update step, after they have been influenced by their parent. The update procedures for the simulations are explicated in the following section.

5   Simulations to Test the Social NAA Assumption

A series of simulations are conducted over various configurations of social networks to test whether the Social NAA assumption holds in all cases on a social level. The Social NAA formalized for this model is the hypothesis that as the total number of hypotheses available in the social network (i.e. here, the number of different hypotheses outputted as a top hypotheses by the agents in their respective H posterior distributions) decreases, the strength in belief in the hypothesis H_{max} (the hypothesis chosen by highest number of agents in the network) should stay the same or increase – i.e. the same number, or more agents should "vote" for H_{max}, whatever it may be, at each time step until convergence when no agent in the network changes their H distribution from the previous time-step. Networks that violate this outcome are counter-arguments against the Social NAA assumption.

In each simulation, the update procedure for each agent at each time step is as follows in two steps:

1. **CALCULATE:**
    i. **IF** at time-step 0
       THEN use the prior distributions for O and E values to calculate the posterior distribution over H using H's conditional probability table, and calculate the marginal probabilities in the standard way [21].
       ELSE (if at time-step 1 or later), update the O variables by aggregating the top hypotheses from the agent's parents' nodes, and set these as the priors for O. If this distribution has changed from the previous time-step, calculate the posterior over H.
    ii. Add the agent's top hypothesis after calculating the posterior to the overall 'vote' distribution for that time-step, to be recorded for later analysis.
2. **UPDATE:** Update the agent's E variables to be equivalent to that of the posteriors over H calculated at Step 1.

Convergence is defined as the time-step at which there is no agent which updates their hypothesis belief posterior H distribution from the previous time-step. In the development of the simulations, it was found a maximum of 10 updates was sufficient to ensure convergence for the network configurations used in this paper. A whole experimental 'run' is therefore considered a set of 10 updates. Each simulation consists of 1000 runs over a network with specified parameters, which are described in the following section.

All simulations were implemented in Python 2.7 and run using iPython notebooks.

5.1   Parameters for Each Simulation: Number of Hypotheses and Agents, Agent Types and Configurations, and Prior Probabilities

Variables that inform the construction of the networks include the number of hypotheses under consideration by the community, the number of agents in the

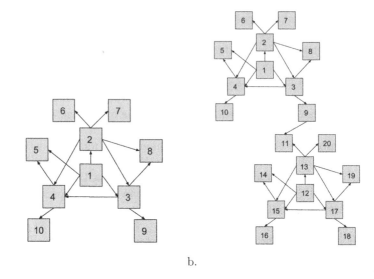

a.   b.

Fig. 4. Networks with 10 (a) and 20 (b) Agents. Position 1 agents are the most connected core agents in the 10-agent networks, and Position 11 agents are bridge agents in the 20-agent networks.

community network, the configuration of the agents in the network structure (i.e. assignment of parent to children agents), and agent types (in terms of the weight they give to their O and E variables in calculating H, and consequently whether they're balanced, mule or sheep agents). The placement of different agent types within the network may also effect whether the Social NAA assumption succeeds—if the most connected and influential agent is a mule type, for instance, the agents directly influenced may never gain access to opinions regarding hypotheses other than the top hypothesis held by the mule type agent.

Six Bayesian network simulations were run using combinations of these variables, which are summarized in Table 1. Networks of 10 and 20 agents were tested, and agents were initialized to have prior beliefs about 3, 4 or 5 hypotheses in the different simulations. Although agents were set to consider a specific number of hypotheses, through various stages in the simulation some agents may not become aware of one or more of them (i.e. being assigned a probability of 0).

The 10 and 20 agent network configurations were adapted from [5] as exemplar of a *core-periphery networks* – that is, several highly interconnected agents are in the network core and less connected agents constitute the network periphery. It serves as an example of a network configuration ubiquitous across a wide variety of communities in reality according to empirical social network literature. Many communities feature networks with a "core-periphery" pattern of a central group or clusters embedded within areas with less dense connections ([5,11] and [19] p. 427). These highly connected clusters tend to include agents with similarities across various domains – that is, "cultural, behavioral, genetic, or material information that flows through networks will tend to be localized"

([19], p. 416). Since it is not computationally possible for way we set up the update step of the Bayesian networks (described above) to allow for bidirectional influence between agents, the core-periphery network structure approximates some well-known dynamics in social networks.

The 10 and 20 agent network configurations are depicted in Fig. 4. The 20-agent network consists of two clusters of 10-agent core-periphery type networks connected by one agent that acts as a 'bridge' between the two clusters such as Agent 11 in Fig. 4 (b) - such a configuration tests whether a bridging agent has a significant influence on introducing the beliefs of one network cluster into the other network cluster.

The types of agents were randomized for each simulation in terms of randomizing the value $W \in [0,1]$ for each agent determining the weight given to O variables. The distribution of the E variable priors were also randomized with each run, and the initial O variable priors were set to a uniform distribution over the set of hypotheses (i.e. $p(O_{Hi} = T) = \frac{1}{n}$ and $p(O_{Hi} = F) = 1 - \frac{1}{n}$ for each hypothesis H_i where there are n hypotheses) before being influenced by the beliefs states of other agents.

It is important to note that in each simulation the most peripheral agents are set as parents to the most core agent, which means that the most core agent is always influenced by a number of agents. This is done in order to accommodate for the lack of bi-directional influence explicitly allowed in Bayesian networks. It is assumed that the most connected agent will also be influenced by a large number of agents due to their large number of parent connections. This stipulation, however, could be omitted in future work to emulate a network that emanates information but does not receive it.

Figure 5 depicts an example of one run from the simulation of ten agents considering four hypotheses. The configuration of the network is shown, as well as changes in the top hypothesis chosen by each agent at each time step in the run. The numbered boxes refer to agents and their place in the network, and the lines with arrowheads represent parent to child influence between agents, with the arrow pointing at the child. The lower the agent's number, the more connected the agent—agent 1, therefore, is the most influential agent in the network, but note it is also influenced by agents 6, 7, 9 and 10 (not depicted to avoid clutter). In this particular run, agent 1 is a balanced type and therefore weighs others opinions and its own preconceived beliefs more or less equally. Each color corresponds to a specific hypothesis, and the color of the agent box indicates the agent's top hypothesis (i.e. the hypothesis with the highest probability). This example depicts a successful run where the social NAA is not violated, because the top hypothesis chosen by most of the agents (hypothesis $H4$ in green) increased in strength (i.e. gained more "votes") with each time step as the number of hypotheses considered viable by each agent decreased. The strongest hypothesis is also the hypothesis originally held and kept by agent 1.

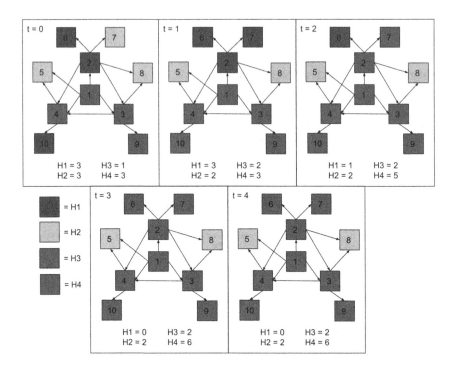

Fig. 5. Example of One Run with Four Updates (First State t = 0). The network converges at time-step 3. (Color figure online)

5.2   Results: The Social NAA Assumption Holds in Most Cases

The results of all six simulations are shown in Table 1. Overall there were violations of the Social NAA assumption in every simulation, showing it cannot be a general rule in social networks, however the number of violations was consistently small—the highest number of violations per simulation was 23/1000 (for the 4-hypothesis 20-agent network), and the lowest number of violations was 6/1000 (for the 3-hypothesis 10-agent network). Therefore, according to this model, in a common network configuration the strength in the top hypothesis usually increases as the number of hypotheses considered feasible by individuals decreases, but this is not always the case.

Further analysis of the the simulations revealed other notable results regarding the differences between agent position and type in violation and non-violation runs. Configurations with violations of the Social NAA assumption tend to feature core (position 1) agents that give less credence to the opinions of others (i.e. are more mule-like) than in non-violation runs—in 5 out of the 6 experiments, the core agent 1 is more mule-like in violating runs than in non-violating runs. In the one exception (3 hypotheses, 10 agents), the difference between the core agent type in the violating and non-violating runs is small, where both could be considered 'balanced' agents. Additionally, in the 20-agent networks the 'bridge' agent (position 11) also tends to be more mule-like in Social NAA assumption violating runs than in non-violating runs.

Table 1. Results of the 6 simulations

Number of hypotheses	Number of agents	Number of violating runs (/1000)	Type of Agent (mean weight of O variables) in Position 1 (violating/non-violating runs)	Type of agent (mean weight of O variables) in Position 11 (violating/non-violating runs)	Steps to convergence (violating/non-violating runs)
3	10	6	**0.533**/0.507	-	2.8/2.8
	20	7	0.221/**0.508**	0.464/**0.492**	3.3/3.5
4	10	13	0.441/**0.506**	-	3.2/3.0
	20	23	0.376/**0.507**	0.300/**0.498**	3.9/3.7
5	10	9	0.401/**0.506**	-	2.8/3.1
	20	14	0.323/**0.493**	0.340/**0.504**	4.3/3.9

These findings suggest that networks featuring particularly influential mule-like agents has an effect on preventing the Social NAA assumption from holding. Intuitively, this is because a mule-like agent that does not change its beliefs in a prominent position prevents the overall dynamics of change of beliefs to swing towards a new consensus immediately, as a sheep-like or balanced influential agent may allow.

Finally, unlike the variables of the type and position of agents in the networks, there was no notable or consistent difference between the number of steps required for the simulation to converge between violation and non-violation runs.

6   Conclusions and Future Work

This paper defined and tested the Social NAA assumption, based on a formulation of the NAA by DHS. The Social NAA assumption was tested by running simulations on a Bayesian network model of interconnected and opinion-sharing agents. The results revealed that the Social NAA assumption was violated in cases where mule-like, or agents less-likely to change their opinions regarding a set of hypotheses, held influential positions in the social network. Nevertheless, the relatively low number of violations of this assumption in a social context indicates that the Social NAA assumption may be a useful heuristic to guide understanding of how individual opinion sharing may influence outcomes of decision-making at the group level. While no strong conclusions can be made at this point, the findings here do support that future work on the relation between the number of perceived feasible options by individuals and group decision-making is warranted. This is the primary contribution of this work: to pose the question of how the relation between the number of choices perceived as feasible by individuals influence choice at the group level, and suggest one approach to answering this question as an initial exploration.

It must be noted that there are significant shortcomings in this model as it stands. First, only a limited number of simulations were conducted. The numerous variables involved in creating the simulations, including configurations of networks, the number of agents, type of agents, and number of hypotheses under consideration, were not fully explored. It is therefore unclear whether the findings from these simulations are merely due to the network configurations chosen rather than other variables such as influences based on agent types and position in the network. In addition, the descriptive results cannot explain whether the noted differences between violation and non-violation runs are significant. In future work, we also intend to take more fine-grained approach to analyzing the probability distributions of the hypotheses rather than just consider the overall vote.

In addition, the model is only capable of representing uni-directional rather than bi-directional opinion sharing between agents. While the network configurations attempted to account for the lack of bi-directional influences by organizing agents in networks found in empirical social network studies, the full extent of opinion sharing between agents was unable to be explored. While the networks were set up to accommodate these limitations, the limitations are nevertheless significant.

The model developed here continues with the Bayesian framework employed by DHS. This is in part to test their assumption within their own framework, and also because multiple hypotheses with probabilities attached to them can easily be represented in Bayesian networks. The extension of previously developed models to further investigate the effects of changes in the number of perceived hypotheses at the group level is sanctioned for future work.

Due to the limitations of Bayesian networks for this research described above, it may be best to adapt existing models of opinion dynamics or belief change to account for multiple hypotheses. Numerous models have been developed to investigate information and knowledge dynamics through communications of agents in a community. Models using dynamic epistemic logic, for instance, are used to model the phenomena of information cascades and pluralistic ignorance (e.g., [1]), while probabilistic (e.g., [8]) and simulation models (e.g., [9,10]) have shown group dynamics of consensus and polarization, among other social processes based on opinion sharing between agents. One direction for future work is to adapt previously these developed models of opinion sharing dynamics to either account for more realistic descriptions of the processes involved in choice ranking and selection at the group level.

Another promising area for future work is exploring the corroborations between the effects of agent types on group decision-making in this model compared to logical models of reasoning about social relationships (e.g. [22]). Logical models of 'threshold influence', for instance, describe how beliefs in community networks are stabilized or destabilized based on configurations of agents and the strength of their influence on others in the network by Liu et al. [17]—both the model developed here and their logical model support the observation that agents' positions in a network are important for the stability of the community's beliefs. An in-depth comparison of these models is therefore an interesting direction for future work.

Finally, another promising direction is the application of the model to real-world data. Since the Social NAA assumption is descriptive, it can be empirically tested to assess whether this model (or another previously mentioned adapted model) can be used to test other relations between the number of choices and group choice. A context that would be well suited for this model is data pertaining to elections, as discussed in Sect. 2. The E variables, which in the current model are under-specified, could be re-conceptualized to capture relevant information about how an agent's preconceived preferences are made. One option is to apply the survey data from the American National Election Studies' (ANES) new 2016 Time Series Study.[6] The survey was given before and after the 2016 U.S. General Election, and includes a variety of questions pertaining to political and policy views, opinions about the candidates (such as a "feeling thermometer" on a 0–100 scale, and emotional responses), and previous voting behavior.

In sum, the purpose of this paper is to offer an initial investigation into how the number of choices perceived as feasible by individual agents in a group setting may influence choice at the group level. We show how one assumption about this relation, the Social NAA, holds in the large majority of simulations in a common social network structure, but there are interesting cases where is does not due to particular opinion-sharing dynamics and the dynamics of particular influential agents in the network. Overall, it is argued that characterizing agents as connected Bayesian networks in a social network, and the relation between the number of choices and group-level choice are directions worth considering in future research.

References

1. Baltag, A., Christoff, Z., Hansen, J.U., Smets, S.: Logical models of informational cascades. Stud. Log. **47**, 405–432 (2013)
2. van Basshuysen, P.: Dawid et al.'s [2015] no alternatives argument: an empiricist note. Kriter. J. Philos. **29**(1), 37–50 (2015)
3. Beach, L.R.: Broadening the definition of decision making: the role of prechoice screening of options. Psychol. Sci. **4**(4), 215–220 (1993)
4. Berlinski, C.: There Is No Alternative: Why Margaret Thatcher Matters. Hachette, London (2011)
5. Borgatti, S.P., Everett, M.G.: Models of core/periphery structures. Soc. Netw. **21**(4), 375–395 (2000)
6. Bovens, L., Hartmann, S.: Bayesian Epistemology. Oxford University Press on Demand, Oxford (2003)
7. Dawid, R., Hartmann, S., Sprenger, J.: The no alternatives argument. Br. J. Philos. Sci. **66**(1), 213–234 (2015)
8. DeGroot, M.H.: Reaching a consensus. J. Am. Stat. Assoc. **69**(345), 118–121 (1974)
9. Douven, I., Riegler, A.: Extending the Hegselmann-Krause model I. Log. J. IGPL **18**(2), 323–335 (2009)
10. Douven, I., Wenmackers, S.: Inference to the best explanation versus Bayes's rule in a social setting. Br. J. Philos. Sci. **68**(2), 535–570 (2015)

---

[6] http://www.electionstudies.org/studypages/anes_timeseries_2016/anes_timeseries_2016.htm.

11. Granovetter, M.: The strength of weak ties: a network theory revisited. Sociol. Theor. **1**, 201–233 (1983)
12. Harman, G.H.: The inference to the best explanation. Philos. Rev. **74**(1), 88–95 (1965)
13. Herzberg, F.: A note on "the no alternatives argument" by Richard Dawid, Stephan Hartmann and Jan Sprenger. Eur. J. Philos. Sci. **4**(3), 375–384 (2014)
14. Kuhn, T.S.: Objectivity, value judgment, and theory choice. In: Bird, A., Ladyman, J. (eds.) Arguing About Science, pp. 74–86. Routledge, New York (1977)
15. Lipton, P.: Inference to the Best Explanation. Routledge, London (2003)
16. List, C.: Social choice theory. In: Zalta, E.N. (ed.) The Stanford Encyclopedia of Philosophy. Metaphysics Research Lab, Stanford University, Winter 2013 Edition (2013)
17. Liu, F., Seligman, J., Girard, P.: Logical dynamics of belief change in the community. Synthese **191**(11), 2403–2431 (2014)
18. Longino, H.E.: Science as Social Knowledge: Values and Objectivity in Scientific Inquiry. Princeton University Press, Princeton (1990)
19. McPherson, M., Smith-Lovin, L., Cook, J.M.: Birds of a feather: homophily in social networks. Annu. Rev. Sociol. **27**(1), 415–444 (2001)
20. Potter, R.E., Beach, L.R.: Decision making when the acceptable options become unavailable. Organ. Behav. Hum. Decis. Process. **57**(3), 468–483 (1994)
21. Russell, S.J., Norvig, P.: Artificial Intelligence: A Modern Approach, 4th edn. Prentice-Hall, Upper Saddle River (2010)
22. Seligman, J., Liu, F., Girard, P.: Logic in the community. In: Banerjee, M., Seth, A. (eds.) ICLA 2011. LNCS (LNAI), vol. 6521, pp. 178–188. Springer, Heidelberg (2011). https://doi.org/10.1007/978-3-642-18026-2_15
23. Shah, A.M., Wolford, G.: Buying behavior as a function of parametric variation of number of choices. Psychol. Sci. **18**(5), 369 (2007)

Readings of Plurals and Common Ground

Kurt Erbach$^{(\boxtimes)}$ and Leda Berio

Heinrich Heine University, Düsseldorf, Germany
{erbach,berio}@uni-duesseldorf.de

Abstract. This paper asks two questions: (i) In an ambiguous context, what is the interpretation of a sentence like *The men wrote musicals*? (ii) How can we succinctly characterize the differences between readings that a sentence has in an ambiguous context, versus readings made available in a specialized context, and those available only because of shared knowledge? While these questions have received much attention, e.g. [1,9–11,20–24,26] i.a., the number of readings such a sentence has in an ambiguous context remains controversial, as is the availability of additional readings, and the means by which speakers become attuned to readings in a given context. To answer the first question we conducted an online study where participants evaluated the truth value of sentences designed to test the meaning of those like *The men wrote musicals*. Results suggest that such sentences get a double cover interpretation (i.e. an interpretation in terms of a relation between sets of individuals, rather than a relation strictly between atomic individuals) in an ambiguous context. We couch these results and the discussion on the availability of other readings in terms of a bipartite Common Ground, where available readings are in the Immediate Common Ground, and other readings can be made available via knowledge in the General Common Ground, thereby answering the second question.

Keywords: Plurals · Salience · Common Ground · Covers · Collectivity · Distributivity · Cumulativity

1 Introduction

The interpretation of sentences with plural nouns like (1) is a controversial topic.

(1)   The men wrote musicals.

For example, Gillon [9–11] argues that plurals are ambiguous, rather than vague or indeterminate, in respect to readings that correspond to minimal covers of the plural noun phrase. Gillon [9] defines a minimal cover as a set that (i) is a subset of the power-set of a set being covered, (ii) contains all of the same individuals as the set being covered, and (iii) contains no set that is a subset of another. This explains why (1) is true of Richard Rodgers, Oscar Hammerstein, and Lorenz Hart, though the men never wrote musicals individually or as a trio,

© Springer-Verlag GmbH Germany, part of Springer Nature 2019
J. Sikos and E. Pacuit (Eds.): ESSLLI 2018, LNCS 11667, pp. 21–41, 2019.
https://doi.org/10.1007/978-3-662-59620-3_2

rather Rodgers wrote musicals with Hammerstein, and he also wrote musicals with Hart. Lasersohn [21,23], however, argues that certain cover readings are never available in certain cases, and that an approach in which plural predicates are ambiguous between collective and distributive interpretations is more sound. Subsequent analyses of plural predicates argue for an additional reading, namely a cumulative reading in which covers of the plural predicates are not specified [1,20,26].

The matter of what readings are available a particular time is discussed at least as a pragmatic issue separate from the semantic question of what readings such sentences can possibly have. Gillon [10,11] assumes that both the beliefs and expectations of interlocutors and the context shape what readings are available. Schwarzschild [24] argues that certain readings must be explicitly mentioned or otherwise salient because of non-linguistic discourse in order to be available. Sternfeld [26] argues that we choose particular readings in effort to make sentences true. In addition to these positions, the question is left open as to the ways in which cognitive mechanisms shape the available readings in a given context.

In this paper, we introduce empirical data from a truth-value judgment task and motivate a new analysis of plural predicates, namely that plural predicates have a double cover interpretation in an ambiguous context—(i.e. an interpretation in terms of a relation between sets of individuals, rather than a relation strictly between atomic individuals or specific sums thereof)—rather than being ambiguous between two or more interpretations. Additionally we specify what it means for readings to be available in a given context by building on the notion of Common Ground [25], and following Krifka [18], we partition Common Ground into parts, namely Intermediate and General Common Ground, also argued for by Berio et al. [2].

2 Background

This paper is focused on sentences like (2), in which there are two plural NPs in a transitive construction that could have a collective, distributive, or any other cover reading.

(2)   Alex, Billy, and Charlie wrote songs.

The collective reading of (2) is such that Alex, Billy, and Charlie all co-wrote the same songs, while a the distributive reading is such that Alex, Billy, and Charlie each wrote their own songs. There are over 100 possible readings of (2), called *cover readings* [9], which are such that some combination of Alex, Billy, and Charlie wrote songs as individuals and/or as groups. For example, one cover of (2) is the reading in which Alex wrote songs individually and with Charlie, while Billy also wrote songs both individually and with Charlie $(a, a \sqcup c, b, b \sqcup c)$. More formally, a cover is a subset of the closure under sum of the atoms a set, and the atoms of the supremum of the subset is equal to the set being covered (where atoms are assumed as countable individuals).

(3)   A covers B iff $A \subseteq {}^*(\text{AT}(B)) \wedge \text{AT}(\sqcup A) = B$

Note that collective and distributive readings are two particular cover readings. In this paper, cover readings other than collective and distributive readings are referred to as *intermediate cover readings*.

   In addition to cover readings, there is a weaker reading called the cumulative reading [1,20,26]. The cumulative reading of a sentence like (4) is one in which three children and five songs were involved in some writing, but it is entirely unclear with respect to which cover of the three children wrote which cover of the songs. In Sternfeld [26], the cumulative reading arises from the ** operation from [19], defined in (5) and results in the logical form in (6).

(4)   Three children wrote five songs.

(5)   For any two-place relation R, let $^{**}R$ be the smallest relation such that $R \subseteq {}^*R$, and
       if $\langle a, b\rangle \in {}^{**}R$ and $\langle c, d\rangle \in {}^{**}R$, then $\langle a \oplus c, b \oplus d\rangle \in {}^{**}R$ [26, p. 304]

(6) $(\exists X)(\textbf{three}(X) \wedge {}^*\textbf{man}(X) \wedge (\exists Y)(\textbf{five}(Y) \wedge {}^*\textbf{songs}(Y) \wedge \langle X, Y\rangle \in {}^{**}\lambda xy[\textbf{write}(x, y)]))$

Sternfeld [26] analyzes sentences like (4) as having a single logical form that automatically generates different readings, including collective, cumulative, and distributive.

   In addition to Sternfeld's [26] analysis, there have been many analyses of sentences with collective/distributive ambiguity, for example [1,9–11,20–24], and there is no consensus on the number of readings available upon the interpretation of sentences like (2) or (4). Analyses can be categorized according to how many readings a sentence with collective/distributive ambiguity might have, e.g. two or many. Though this categorization does not capture the subtleties that make each theory different from one another, it highlights the underlying point: the interpretation of sentences like (2) remains a matter of debate. While all agree that the sentences in question can have many logically possible readings, two open questions remain: (i) What is the interpretation of such sentences an ambiguous context? By this we mean, in an ambiguous context, what is it that native speakers of English understand such sentences to mean, and how many readings are part of this meaning? (ii) In a context where such a sentence is ambiguous between two or more readings, how is it that these readings are "available" (or in terms of Schwarzschild [24], "contextually salient"), and what does it mean for certain cover readings to never be available as argued by Lasersohn [21,23]?

2.1   Previous Analyses

Two-Reading Analyses. Two-reading analyses include Lasersohn [23], who argues that sentences with both collective and distributive readings are straightforward examples of ambiguity. He argues for this analysis by showing that it must be the case that both collective and distributive readings are available for certain pairs of sentences to be true. (7-a) and (7-b) are one such pair which can both be true at the same time only if both sentences are ambiguous between collective and distributive readings.

(7)   a.   John and Mary earned exactly \$10,000.
        b.   John and Mary earned exactly \$5,000.   [23, p. 131]

In other words, (7-a) and (7-b) can both be true because the collective reading of (7-a) has the same truth conditions as the distributive reading of (7-b). Lasersohn [23] argues that, without assuming these two readings are straightforwardly available, (7-a) and (7-b) cannot both be true at the same time, which is not the case.

Lasersohn [21] argues against multi-reading analyses—e.g. Gillon [9]–claiming that certain minimal cover readings are never available. For example, under Gillon's [9] analysis, (8) is incorrectly predicted to be a true statement when John, Mary, and Bill are teaching assistants (TAs) who each made exactly \$7,000 every time they occurred as a member of co-TA teams: {John and Mary}, and {John and Bill}. In other words, each pair of John, Mary, and Bill collectively earned exactly \$14,000, so (8) should be true but is not according to Lasersohn [21].

(8)   The TAs were paid exactly \$14,000 last year.   [21, p. 131]

Because Lasersohn [21] assumes that (8) is false in a situation where the pairs {John and Mary}, and {John and Bill} each made exactly \$14,000, a multi-reading analysis that includes minimal covers like Gillon's [9–11] is untenable. This argument, in addition to the argument that sentences like (7-a) and (7-b) must have two readings in order to both be true, is why Lasersohn [21] analyzes such sentences as straightforwardly ambiguous between only collective and distributive readings.

Many-Reading Analyses. Many-reading analyses include Gillon [9–11], Sternfeld [26], Beck and Sauerland [1], and Landman [20]. In Gillon's early work, [9], he argues that sentences like (2) are ambiguous in respect to their truth conditions, which is a set of minimal covers—i.e. sets of subsets of pluralities, in which none of the subsets overlap with the union of the others, and the union of all subsets is equal to the plurality itself (9).

(9)   A minimally covers B iff A covers B $\wedge \neg \exists X(X \subseteq A \wedge \bigcup (A\text{-}X)$ covers B)

In other words, (2) has eight possible interpretations which correspond to the minimal covers of the subject NP. As discussed above, if we are discussing Rodgers, Hammerstein, and Hart, the sentence, *The men wrote musicals*, is true under the reading where Rogers and Hammerstein wrote musicals together as did Rogers and Hart, though it is false under both distributive and collective readings.

More recently, Gillon [10] has argued that extra-grammatical conditions constrain the possible readings of a given plurality. He insists that context can make available intermediate minimal cover readings—i.e. minimal cover readings other than collective and distributive. Gillon [11] gives (10-a) as an example of a context that makes intermediate cover interpretations available.

(10)   a.   A chemistry department has two teaching assistants for each of its courses, one for the recitation section and one for the lab section. The department has more than two teaching assistants and it has set aside $14,000 for each course with teaching assistants. The total amount of money disbursed for them, then is greater than $14,000. At the same time, since the workload for teaching a course's section can vary from one section to another, the department permits each team of assistants for a course to decide for itself how to divide the $14,000 the team is to receive.

    b.   The T.A.'s were paid their $14,000 last year.   [11, p. 483].

While (10-a) does not explicitly point to which minimal cover is true, it nevertheless gives the context necessary to know that distributive or collective readings of (10-b) are not true, and an intermediate cover reading is necessary. With (10-a) as context, (10-b) is straightforwardly true, according to Gillon [11]. In addition to explicit contextual information, Gillon [11] assumes that the beliefs and expectations of interlocutors have a role in determining the available readings of sentences. Gillon [11] argues that a sentence like *The man surrounded the town* is grammatical and true in a novel like *Gulliver's Travels* where the readers know that Gulliver is such a size that he can indeed surround an entire town. For Gillon [11], it is extra-grammatical information such as this that restrict the possible readings.

Sternfeld [26], Beck and Sauerland [1], and Landman [20] each provide different multi-reading analyses than Gillon, each arguing for the existence of the cumulative interpretation [1, 20, 26]. With the addition of the cumulative interpretation, *The men wrote musicals* is not only true of the intermediate cover that specifies the pairs {Rogers and Hammerstein} and {Rogers and Hart}, but it is also true of the weaker, cumulative reading in which neither the cover of Rodgers, Hammerstein, and Hart nor the cover of musicals is specified with respect to the writing. Sternfeld [26] also argues that an interlocutor's desire to find a true reading of a particular sentence determines which reading a sentence is chosen to have in a particular interpretation, therefore it is the truth conditions of a sentence that dictate the readings in a particular context.

Hybrid Analyses. Schwarzschild [24] argues for an analysis that incorporates elements of both two- and a many-reading analysis. In his context based analysis, he analyzes plural predicates as having a single meaning that can be indexed to any cover reading in the appropriate context. According to Schwarzschild, [24], "whether or not a certain intermediate reading is available seems to have to do with the context not with the semantics of particular lexical items" (p. 66). He therefore proposes the following generalization to account for cover readings:

(11)   [$_S$NP$_{plural}$ VP] is true in some context Q iff there is a cover C of the plurality P denoted by NP which is salient in Q and VP is true for every element in C.

While Schwarzschild [24] does not explicitly argue for cumulative readings, Sternfeld [26] uses definitions such as (11) as the formulation of cumulative readings. Cumulative readings are therefore built into Schwarzschild's [24] analysis although never evoked.

The generalization for distributive readings is formalized in (12), where Part is the one place distributivity operator and Cov is free variable over sets of sets of the domain of quantification, the value of which is determined by the linguistic and non-linguistic context.

(12) $x \in \|Part(Cov)(\alpha)\|$ if and only if $\forall y[(y \in \|Cov\| \wedge y \subseteq x) \rightarrow y \in \|\alpha\|]$
                                                                    [24, p.71]

Schwarzschild [24] specifies the translation rule in (13) which means that a plural predicate is indexed to a particular cover reading.

(13)   Plural VP rule:
         If α is a singular VP with translation α', then for any index i, $Part(Cov_i)(\alpha')$ is a translation for the corresponding plural VP.

These rules allow any cover reading to be indexed given the right context. (14-a), for example, therefore has the logical form in (14-b), where the two-place Part operation distributes the predicate to the subsets of the indexed cover(s), Cov_i.

(14)   a.   The musicians wrote songs.
         b. $(Part(Cov_i)(wrote'))(songs')(the\text{-}musicians')$

Schwarzschild [24] concludes that the absence or presence of a given cover interpretation depends, to some extent, on the same sorts of things that other pragmatic phenomena like anaphoric reference depend on, like salience. However, in an ambiguous context, collective and distributive readings are made salient by the plural noun phrase itself. So, while plural predicates have a single interpretation, it is in some sense a place-holder for one or more indexed cover readings.

The discussion of salience in Schwarzschild [24] builds on work on anaphoric reference with pronouns. He notes that, for pronouns, it is necessary for the referent to be explicitly mentioned in order to be accessible, though this is not a sufficient condition. These points are respectively reflected in (15) and (16).

(15)   Nine of the ten balls are in the bag. It's under the couch.   [24, p. 94]

(16)   The boys and the girls entered the room (separately). They were wearing hats and they were wearing skirts.   [24, p. 95]

In (15), the first sentence indicates there is a ball not in the bag, though because this ball is not explicitly mentioned, the follow-up sentence seems odd: the referent of *it* is not entirely clear. This is taken to indicate that referents of pronouns must be explicitly mentioned. In (16), though two groups of children are explicitly mentioned by distinguishing them according to gender, the sequential uses of *they* in the follow-up sentence does not straightforwardly correspond to respective anaphoric reference. This mismatch of explicitly mentioned anaphora is taken to indicate that their explicit mention is not a sufficient condition for anaphoric reference. Schwarzschild [24] extends this analysis to cover readings, arguing that (16) *the boys and the girls* is explicit mention of a cover—the sum of the set denoted by *the boys* and the set denoted by *the girls* covers the implicit NP *the children*—and therefore that the sort of pragmatic and extra-linguistic principles that apply to pronominal anaphoric reference also apply to cover interpretations.

2.2   Availability and Salience

The question that remains, is how can we succinctly characterize the differences between readings that an a sentence has in an ambiguous context, versus readings made available in a specialized context, and those available only because of shared knowledge? Schwarzschild's [24] position on the explicit salience of a cover is stronger than Gillon's [10, 11] less-clear notion of context shaping the domain of quantification. For example, it is not clear if Schwarzschild [24] would agree that the intended minimal cover has been made salient in the context Gillon [11] provides in (10-a). Similarly, Sternfeld's [26] position on the goal of interlocutors to find a true reading is stronger than Gillon's [10, 11] assumption that the beliefs and expectations of interlocutors and shape what readings are available. For example, Sternfeld [26] might (not) argue that an interlocutor does not need the rich context provided in the novel *Gulliver's Travels* to find a true reading of the sentence *The man surrounded the town*. At the same time, there has been no discussion of Lasersohn's [23] position that certain readings are never available, and it is unclear if there is any extent to which such a position is compatible with the others. These potential incompatibilities make clear the fact that there is no straightforward characterization of what it means for a reading to be available in a given context.

Contexts of Quantification. A related set of questions regards the inquiry about domain quantification and context carried on in the context of Philosophy of Language. In what follows, we will argue that, while that debate adds an interesting perspective to the problem raised by ambiguous sentences with plural predicates, it does not exhaust all the pending questions.

A relevant analysis of context and quantifiers is that by Bianchi [3], where the debate about intentional and objective context is connected to the conditions for determining the quantification debate. Bianchi [3] follows the analysis of Gauker [7] in distinguishing between how theories predict a restriction of the domain of quantification. These theories fall into one of two classes, those with Intentional Perspective on Context (IPC) and those with Objective Perspective on Context (OPC) [7]. The main intuition underlying IPC (Intentional Perspective on Context), is that the intention of the speaker counts towards determining the truth conditions of a proposition, thereby, following the direction of the Gricean tradition [12,13]. OPC (Objective Perspective on Context) argues that the propositional content of the utterance is to be determined by looking at the objective features of the context and that the intention of the speaker is irrelevant for determining the semantic value. As Bianchi [3] puts it, the debate is connected to the more general question regarding what determines reference in cases, like indexicals and demonstratives, where contextual information has to be taken into account [16,17].

Gauker [7] argues against IPC with examples like (17), which, in certain contexts, are infelicitous despite the intention of the speaker. The context is the following: Scout and Jo are playing with Jo's marbles in her room. Scout utters

(17)   All of the red marbles are mine!

With (17), Scout intends to refer to the marbles that are laying in her own room, under the bed, away from the current communicative situation involving Jo. Gauker [7] uses a similar example to argue against IPC, given the fact that, if the speaker's intention was relevant to determine the propositional content of (17), we would be forced to consider the proposition to be true, even though, according to our intuition, it is more likely to be false; Jo could rightly be upset with Scout, thinking she wants to claim property of her red marbles. Gauker [7] then, openly criticizes IPC because it appeals to intentions and mental representations, arguing that what is relevant for determining the domain of quantification is, instead, the objective context of the utterance–i.e, the relevant states of affairs in the world.

Bianchi [3] defends IPC from Gauker's [7] attacks by arguing that IPC does not simply defend a view according to which *any* semantically relevant intention of the speaker related to the communicative situation is relevant for individuating the quantification domain; instead, the only intentions that are to be considered are those made *available* in respect what she calls an *availability constraint*. In other words, an intention that is made available to the addressee, and that therefore is "non arbitrary—that is connected with a particular external context, or a suitable behavior, or else an appropriate co–text, that enable the addressee to determine the referent" [3, p. 389]. Such an intention is recognized on the basis of the physical surroundings (or "external" facts), linguistic co-text, and background knowledge.

Pending Questions in the Analysis. Bianchi's [3] solution, we would argue, captures a relevant issue, namely that there is a plurality of factors involved in determining the relevant proposition expressed by an utterance like (17). With respect to the readings of sentences with plural predicates, following Bianchi [3], the speaker may have a particular reading of the sentence in mind, however, without the speaker making the particular reading explicitly available with (extra-)linguistic information in the context, the particular reading will in no way be available to hearer. In other words, a sufficient condition for making a particular reading available is for the speaker to make it known to the hearer that a particular reading is intended. Not only does this clarify what it means for a reading to be made contextually salient via (extra-)linguistic information, but it also points to what it means for a reading not to be available, namely that it is not brought into focus by the speaker nor the extra-linguistic information in the context. In this sense, Bianchi points out a relevant fact about the communicative situation above: shared knowledge is relevant to make the communicative context successful. However, her focus is on defining the relevant features of the context that grant the truth conditions and, in this sense, her account does not point to how the sentence is interpreted in an ambiguous context, all the ways in which the sentence could potentially be understood, or the relation between the semantic content of the sentence and its speakers' interpretation.

3   Main Data

While Bianchi's [3] contribution points out the role of a shared intention in context, the interpretation of a sentence like (2) in an ambiguous context is still unclear. In addition to distributive and collective readings of plural predicates, lexical modifiers like *each* have a distributive effect, and modifiers like *together* have a collectivizing effect [10,24,27]. These lexical modifiers can therefore be used to restrict the possible interpretations to distributive, (18-a), or collective, (18-b).

(18)   a.   Alex and Billie wrote songs individually.
         b.   Alex and Billie wrote songs together.

If plural predicates like *wrote songs* have all minimal cover readings available as argued by Gillon [9] in his early work, then (2) should be equally ambiguous in respect to the combinations of song-writers listed in (19).

(2)   Alex, Billie, and Charlie wrote songs.

(19)   a. $a \sqcup b \sqcup c$ e. $c, a \sqcup b$
         b. $a \sqcup c, b \sqcup c$ f. $b, a \sqcup c$
         c. $a \sqcup b, b \sqcup c$ g. $a, b \sqcup c$
         d. $a \sqcup b, a \sqcup c$ h. a, b, c

If all minimal cover readings are equally available, then it should be possible to refer to a subset of the minimal covers by adding lexical modifications. For example, (20-a) is true of a set of minimal covers, and (20-b) is true of a subset of those minimal covers.

(20)   a.   Alex, Billie, and Charlie went to the music studio. The musicians wrote songs.
         b.   Alex and Billie didn't write songs individually.

The set of minimal covers that could be true of both (20-a) and (20-b) includes the distributive interpretation and every other cover in which the predicate distributes to either Alex or Billie individually.[1] The only available interpretations would be those in which Alex and Billie are part of a collective interpretation. The potentially true minimal covers are listed in (21), along with the false minimal covers, which are crossed out.

(21) a. $a \sqcup b \sqcup c$ e. $c, a \sqcup b$
     b. $a \sqcup c, b \sqcup c$ f. ~~$b, a \sqcup c$~~
     c. $a \sqcup b, b \sqcup c$ g. ~~$a, b \sqcup c$~~
     d. $a \sqcup b, a \sqcup c$ h. ~~a, b, c~~

It is also possible to use modifiers to eliminate collective interpretations for particular individuals. In (22) for example, the use of *together* in (22-b) negates the scenarios in which Alex and Billie are predicated over collectively.

(22) a. Alex, Billie, and Charlie went to the music studio. The musicians wrote songs.
     b. Alex and Billie didn't write songs together.

The set true and false minimal covers for (22-a) and (22-b) are listed in (23)[2].

(23) a. ~~$a \sqcup b \sqcup c$~~   e. ~~$c, a \sqcup b$~~
     b. $a \sqcup c, b \sqcup c$ f. $b, a \sqcup c$
     c. ~~$a \sqcup b, b \sqcup c$~~   g. $a, b \sqcup c$
     d. ~~$a \sqcup b, a \sqcup c$~~   h. a, b, c

Taking these modifications one step further, only a single minimal cover is available when using both *individually* and *together* in the same sentence. For example, given (24-a) as a context, (24-b) negates all minimal covers in which *wrote songs* gets a collective or distributive interpretation in respect to Alex and Billie.

(24) a. Alex, Billie, and Charlie went to the music studio. The musicians wrote songs.
     b. Alex and Billie didn't write songs individually or together.

Both (24-a) and (24-b) are true if Alex and Charlie wrote songs together and Billie and Charlie also wrote songs together. The true and false minimal covers of these two sentences are listed in (25).

---

[1] While it could be the case that the use of *The musicians* as opposed to the plural pronoun could be taken as an indication that one of Alex, Billie, and Charlie is not a musician, we contend that these sentences still allow for the reading in which Alex, Billie, and Charlie are all musicians.

[2] Though $p \sqcup q$ is only a subpart of $p \sqcup q \sqcup r$, this reading is assumed to be canceled via implicature.

(25)   a.   ~~$a \sqcup b \sqcup e$~~   e.   ~~$e, a \sqcup b$~~
       b. $a \sqcup c, b \sqcup c$ f.   ~~$b, a \sqcup e$~~
       c.   ~~$a \sqcup b, b \sqcup e$~~   g.   ~~$a, b \sqcup e$~~
       d.   ~~$a \sqcup b, a \sqcup e$~~   h.   ~~a, b, e~~

Experimental Design. An empirical study was designed to test the interpretations of the sets of sentences, like those in (20), (22), and (24). A truth-value-judgment survey was conducted with 32 native English speakers through Prolific.ac. The participants were presented with 45 test items containing a set-up like (24-a) and a follow-up like (24-b). Participants were told to judge whether the follow-up sentence could be true or must be false in respect to the set-up preceding it. While these directions were written above every set of sentences, the options the participants clicked on were simply labeled *True* and *False*. The 45 test items exemplified one of the three conditions in (20), (22), and (24): 15 test follow-up items contained *individually*, 15 contained *together*, and 15 contained both *individually* and *together*. While each set-up contained a subject DP with three individuals, they varied with respect to the names of the individuals and the VP that followed, but all set-up sentences could be interpreted as true of any cover reading. Participants were also asked to judged the truth value of 45 filler items that could be true or must be false depending on their lexical modifiers. The total number of items expected to be true or false was equal. Because the pairs of sentences were presented to participants in this experimental context, the context is assumed to be ambiguous, and without any indication that particular readings should (not) be available. We therefore take the judgment of the follow-up sentence as indication of what readings are available of the sentences with plural predicates in the set-up in the manner described above with examples (20)–(24). If all of these follow-up sentences are judged to be possibly true, then it could be the case that the plural predicates are straightforwardly ambiguous between all minimal cover interpretations as argued in Gillon's [9, 10]. Second, if (20) and (22) are judged to be possibly true, and (24) is judged to be necessarily false, then plural predicates have distributive and collective readings in an ambiguous context but intermediate cover readings are not available, as argued by Lasersohn [23] and [24]. Alternatively, if all follow-up sentences are judged to be false, then it is the case that only the cumulative interpretation is available in an ambiguous context, and all other interpretations are derived or indexed.

Results. The results of the study show that there is a significant difference in the way that the truth of sentences with both *individually* and *together* are judged relative to sentences with only one of the two lexical modifiers. Using a binary logistic regression model (lme4 package in R), and the conditions and judgments as arguments, the judgments of test condition with both *individually* and *together* were found to be significantly different ($p < 0.001$) than judgments of the condition in which sentences only contained *together* as a lexical modifier.

Sentences that only contained *individually* as a lexical modifier were found to be judged no differently ($p = 0.282$) than those that only contained *together*. These results show that despite the fact that each follow up sentence is true in respect to its preceding context, speakers do not judge sentences in the test condition to be true at the same rate at which they judge sentences in the other conditions to be true.

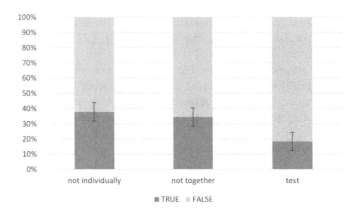

Fig. 1. Average percentage of true and false judgments by condition

The average percentage of true and false judgments for sentences in each condition is presented in Fig. 1. This graph shows that follow up sentences with only one of the two lexical modifiers are judged as necessarily false a majority of the time, while follow up sentences with both lexical modifiers are judged as false an even larger majority of the time. In other words, negated follow up sentences that restrict the set of true minimal covers with the lexical modifiers *individually* or *together* are generally judged to be false.[3] This is a surprising result given the plural predicates are said to have both collective and distributive readings, yet neither reading seems to be available when the subjects were asked to interpret the possible truth of follow-up sentences. If the collective reading was available, then the follow-up sentences negating the distributive reading should all have been true. Furthermore, if the distributive reading was available, then the follow-up sentences negating the collective reading should have been true.

Discussion. We take the fact that the follow-up sentences were judged to be false to suggest that the plural predicate they follow is not straightforwardly ambiguous between all minimal covers as argued for by Gillon's earlier work [9,10]. It also does not seem to definitely be the case that, in this context, they are ambiguous between collective and distributive interpretations argued by

---

[3] Although conditions negating just one of the collective and distributive readings respectively are close to 50%, they are significantly different ($p < 0.001$) than an artificial data set in which the same number of items were equally split between true and false judgments.

Lasersohn [23] and Schwarzschild [24][4]. Instead of any of the aforementioned analyses, the empirical data seems to point toward an analysis in which neither the distributive, collective, nor intermediate cover readings are available readings of such sentences in an ambiguous context.

4 Analysis

The results of the study point towards an analysis in which sentences like *The men wrote songs* have only a weak interpretation in an ambiguous context, but can also index other readings when they are made salient. Because none of the aforementioned analyses argue for such an analysis, we build our analysis on that of Landman [20], whose "double cover interpretation", from which minimal cover interpretations can be derived, expresses a relation between sets of individuals rather than a relation strictly between atomic individuals. This weaker form of a cumulative reading is advantageous because it allows for a more straightforward relationship to intermediate cover readings, which, by nature, involve sets of individuals rather than a relation strictly between atomic individuals.

For Landman [20], intermediate cover interpretations are the result of a special contextual mechanism that weakens the interpretations of verbs. In respect to a plural argument like *the musicians* that denotes three individuals, Alex, Billy, and Charlie, or $a \sqcup b \sqcup c$, a minimal cover like Alex and Charlie, and Billie and Charlie ($a \sqcup c$, $b \sqcup c$ in (26)), can be the agent of a plural predicate, e.g. (27)[5], so long as one has a definition of cover roles (28), a definition of covers (29), and a type shifting principle for verbs that allows verbs with plural roles to be turned into cover roles (30).

(26) $\{a \sqcup c, b \sqcup c\} \in$ *MUSICIAN
      [[the musicians]] $= \sigma(\text{*MUSICIAN}) = \sqcup\{a \sqcup c, b \sqcup c\} = a \sqcup b \sqcup c$

(27) [[*The musicians wrote songs*]] $= \begin{cases} \exists e \in \text{*WRITE}: \\ a \sqcup b \sqcup c = \sigma(\text{*MUSICIAN}) \wedge \\ {}^C\text{Ag}(e) = \uparrow(a \sqcup b \sqcup c) \wedge \\ \exists y \in \text{*SONG} \wedge {}^C\text{Th}(e) = \uparrow(y) \end{cases}$

---

[4] Among the possible interpretations of the results, one might argue that the presence of negation in the follow-up sentences might be the reason why the participants judged them to be false—i.e. negative sentences could have made the parsing harder and the judgment more difficult. We thank the anonymous reviewer for the observation. However, to disambiguate between alternative analyses—two-reading analyses versus many-reading analyses—it was necessary to have a negation in the test sentence since the positive equivalent would not distinguish between the two alternatives. In order to make the control sentences comparable, it was sensible to keep negation in all the sentences, to avoid a result biased by the presence of negation in the test sentences but not in the control ones. In this way, the difference between response rates for control sentences and test sentences is not attributable to the presence or absence of negation.

[5] AT(d) is the set of atoms below d: if $d \in D$ then $AT(d) = \{a \in AT: a \sqsupseteq d\}$.

(28) Let R be a thematic role
^{C}R, the cover role based on R,
is the partial function from D_e to D_d defined by:
^{C}R(e) = a iff a $\in$ ATOM $\wedge \sqcup(\{\downarrow$ (d) $\in$ SUM: d $\in$ AT(*R(e))\}) $= \ \downarrow$(a)
undefined otherwise   [6, p. 210]

(29) group β is a subgroup of α iff $\downarrow$ (β) $\sqsupseteq \ \downarrow$ (α).
Let X be a set of subgroups in group α.
X covers α iff $\sqcup \{\downarrow$ (x) $\in$ X\} $=\downarrow$ (α)   [6, p. 211]

(30)
$\lambda x_n...\lambda x...\lambda x_1.\{e \in {}^*V:...{}^*R(e)=x...\} \rightarrow$
$\lambda x_n...\lambda x...\lambda x_1.\{e \in {}^*V:...{}^CR(e)=x...\}$ [6, p. 211]

For Landman [20], cover readings are those in which there are plural agents
of sums of events. Such readings are made possible by cover roles, which are
defined in (28). If the plural role R has atoms d, and those atoms can be type-
shifted down with the operation $\downarrow$, and we can take the sum of those type-
shifted individuals, and that sum of type-shifted individuals is equal to the plural
individual made from the group a, then a is a cover role. More plainly, if the
agent of an event is a sum of groups, then that agent is a cover role. This is
exactly what occurs when a sentence like (31-a) is used to describe the event
that is described in (31-b)—i.e. an event in which $a \sqcup c$ and $b \sqcup c$ are the agents
of separate song writing events.

(31)   a.   The musicians wrote songs.
        b.   Alex and Charlie wrote songs together, and Billie and Charlie wrote
              songs together.

In order to derive the interpretation in (27) from that of (31-b), the following
must occur: $\uparrow (a \sqcup c)$ and $\uparrow (b \sqcup c)$ must be group atoms (made via the type
shifting operation $\uparrow$[6]) that are the agents of events e and f respectively (32).

(32) $\uparrow (a \sqcup c)$ = Ag(e)
        $\uparrow (b \sqcup c)$ = Ag(f).

The plural agent of the sum of events e and f is equivalent to the sum of the
groups $\uparrow (a \sqcup c)$ and $\uparrow (b \sqcup c)$:

(33) *Ag(e$\sqcup$f) = $\uparrow (a \sqcup c) \sqcup \uparrow (b \sqcup c)$ [20, p. 212]

The set of atoms below the plural agent in (33) is the set containing the two
groups $\uparrow (a \sqcup c)$ and $\uparrow (b \sqcup c)$:

(34)   AT(*Ag(e$\sqcup$f)) = $\{\uparrow (a \sqcup c), \uparrow (b \sqcup c)\}$ [20, p. 212]

Given the definition of cover roles, (28), it is possible to take the closure under
sum of the set of atoms below the plural agent, and therefore get the supremum

---

[6] One function of the type shifting operation $\uparrow$ is to turn plural individuals into group
atoms; see [20] for details.

of the groups of agents (35), which upshifted, is equivalent to the plural agent of events e and f (36).

(35) $\sqcup\{\downarrow(d): d \in AT(*Ag(e \sqcup f))\} = \sqcup\{a \sqcup c, b \sqcup c\} = a \sqcup b \sqcup c$

(36) $*Ag(e \sqcup f) = \uparrow(a \sqcup b \sqcup c)$

The type-shifting principle for verbs, (30), allows the meaning of the verb *write* to be shifted cover interpretations:

(37) $write \rightarrow \lambda y \lambda x.\{e \in *WRITE:^C Ag(e)=x \wedge {}^C Th(e)=y\}$

This derivation provides a cover agent for the interpretation of (27) from the interpretation of (31-b).

   While Landman [20] provides this mechanism for building plural predicates from covers, he argues that these are special cases that are not part of the interpretation of the verb. He argues that the sentences in question have four scopeless readings (double collective, collective-distributive, distributive-collective, and double-distributive–i.e. cumulative) if plural noun phrases fill the roles of the verb, and five other readings are available depending on how a particular scope mechanism is invoked. The cumulative interpretation is relational–i.e. it is not a statement about each individual denoted by the arguments of a transitive verb, and it is not about a predicate and one argument: it is about the relation between the predicate and its arguments. The cumulative reading (31-a) indicates that (i) there is a set of musicians, (ii) there is a set of songs, (iii) every one of the musicians wrote at least one of the songs, and (iv) every song was written by one or more of the musicians. The cumulative interpretation can be type-shifted to the "double cover interpretation", from which minimal cover interpretations can be derived, meaning that a relation between subgroups is expressed rather than a relation between individuals.

   Adopting on the idea of Schwarzschild [24] that a plural predicate has one meaning that can index cover interpretations, and also the idea from Landman [20] that cover readings are derived from a double cover interpretation, we motivate an analysis in which plural predicates have a single, general interpretation from which all cover interpretations are indexed. The double cover reading from Landman [20] provides a weak, general meaning for the plural predicate, and by adding indexing, specific interpretations can be salient. The required translation entails the following rule.

(38)   If α is a singular transitive verb phrase with translation A, then for any index i, $\exists e \in *A : {}^{C_i} Ag(e) = x \wedge {}^{C_i} Th(e) = (y)$ is the translation for the corresponding plural transitive verb phrase.

If a particular cover is not indexed in the context–i.e. the index is left unspecified as i–then the plural predicate is straightforwardly interpreted as a dual cover reading. The reading indicates (i) that there is a sum of writing events, (ii) there is a sum of groups of musicians (Alex, Billie, and Charlie in (20), (22), and (24)) as a plural agent, (iii) there is a sum of groups of songs as a plural theme:

$$(33) \quad \llbracket \textit{The musicians wrote songs} \rrbracket = \begin{cases} \exists e \in {}^{*}\text{WRITE}: \\ a \sqcup b \sqcup c = \sigma({}^{*}\text{MUSICIAN}) \land \\ {}^{C_i}\text{Ag(e)} = \uparrow (a \sqcup b \sqcup c) \land \\ \exists y \in {}^{*}\text{SONG} \land {}^{C_i}\text{Th(e)} = \uparrow (y) \end{cases}$$

While this seems very similar to a distributive interpretation (and in Landman's [20] framework, the double cover interpretation is a type-shifted double-distributive (cumulative) interpretation), without indexing a particular cover, it is impossible to tell exactly which (covers of) musicians wrote exactly which (covers of) songs. It is therefore distinct from Landman's [20] scoped distributive readings where the set of musicians would necessarily distribute to either distinct sets of songs, or the same set of songs.

4.1 Interpretation in Context: Quantification and Common Ground

While we have provided a semantic analysis of the interpretation of sentences like *The men wrote musicals*, it is still unclear what it means for a reading to be indexable in context given Gillon [10,11], Schwarzschild [24], and Sternfeld [26] all provide different notions of what it means for such readings to be contextually available. Bianchi's [3] availability constraint provides a possible replacement for the notion of being contextually available and to the IPC vs. OPC debate in suggesting that communicative intentions have to be made available and readable in the objective context in order to help determine the content of the proposition. This solution, however, does not provide any indication of the relationship between the reading of a sentence in an ambiguous context and the reading(s) that may or may not be indexed in a given context.

To capture these relationships, we invoke a bipartie notion of Common Ground [25], as first called for by Krifka [18]. Berio et al. [2], modify Stalnaker's [25] classic notion of Common Ground by distinguishing between ICG (Immediate Common Ground) and GCG (General Common Ground). ICG and GCG are different levels of shared information that are both involved in conversation, though they are different in terms of their relation with the communication at hand. GCG contains semantic and world knowledge, and social and linguistic practices, which are stored in long term memory but not constantly recalled/activated in conversation. In ICG, discourse-specific referents are brought into focus by linguistic and perceptual mechanisms.

GCG includes knowledge about language in general, idiolects that are used in different linguistic groups, conventional implicatures, and conventions in general. Information such as encyclopedic entries, for instance, or definitions, are included in the GCG, and so are the principles and knowledge of logic that are shared among individuals in a given community. However, not only propositional and linguistic information is involved in the GCG; on the contrary, perceptual information in form of memory traces and recollection can be found at the same level of CG, since they pertain to information that is stored in the long term memory—e.g. we can refer in conversation to a dog we saw earlier in the day. While the way we refer to the dog and the fact that we will bring it into the conversational focus is a matter of ICG, the fact that we have seen it is likely to be stored in episodic memory format. Note that this account of Common Ground is to be considered along the lines of the proposals made by Clark and Horton, [4,6,14], to describe shared information among conversation participants in terms of grounded cognitive mechanisms and to get away from the presuppositional account of Common Ground of the Stalnakarian tradition [25].

ICG is specific to communicative situations at hand, and includes linguistic and non-linguistic information. This is the level where triple co-presence as defined by [4,6] is mostly relevant, i.e. the shared and joint attention of, for instance, two actors on an object. Tomasello [28] invokes a similar notion for language acquisition called "shared attentional frame". Co-presence does not have to be physical, but can be linguistic as well; in that case, joint attention can be focused through linguistic reference. Fundamentally, not everything that belongs to the shared immediate context is part of the speakers' ICG, since many things can be part of the environment, linguistic or non-linguistic, without being salient or attended to by the speakers. ICG, in this sense, is the level where elements like perceptual salience play a relevant role [5]. The ICG can be very rich at times, and very poor in other situations. This account predicts that, when the amount of information shared in the specific situation is minimal, the speakers will rely on information they share on a general level, opting possibly for what is the most frequent interpretation or, eventually, engaging in some strategic thinking to assess what is indeed shared in the situation (as in analyses of Common Ground and memory processes like [14,15]).

The difference between ICG and GCG can be conceptualized in terms of which kind of memory is involved; while information and knowledge regarding word use and meaning and information about the world is a matter of General Common Ground and it involves long term memory and working memory, what is shared in the Immediate Common Ground is mostly a matter of working memory, as it entails the ability to keep track of contextual clues, of both linguistic and non-linguistic nature. In the same way working memory and long-term memory constantly interact, the interaction between ICG and GCG is constant and dynamic. This is based on what is commonly called *memory resonance*, i.e. the parallel elaboration of cues in working memory with stored, long-term memory information [8]. Horton and Gerrig [14,15] make such an appeal to memory resonance to explain interaction at Common Ground, however, there is a focus on specific cases of shared information, i.e. cases of past physical co-presence.

As an illustration of the interaction between ICG and GCG, consider the following example. When speaking to another adult who is fluent in English, I can assume shared knowledge (GCG) of, say, the kind of entity and animal that an elephant is. If the interlocutor knows me and my office well enough, the fact that I have elephant figurines on my desk is also in our GCG. If I am in my office, speaking over the phone to this interlocutor, then I will have to rely on linguistic cues to invoke the view of my desk they would have if they were there in person, or on my linguistic action of referring to elephants (ICG) to retrieve the information about figurines on my desk from the GCG, so my interlocutor will know I am not referring to an actual animal if I utter something like *An elephant fell*. In such a situation it will have become a matter of ICG that the elephant is a figurine, that it is perceived by both of us, and so on. Note that, if the person is in the office with me, the simple presence of the elephants on my desk will not be sufficient for them to be part of the Immediate Common Ground with my interlocutor. What it will be necessary is for the elephants to be salient enough in the communicative situation, for example by me referring to them with linguistic or non-linguistic means.

Building on Berio et al. [2], we rely on a domain of quantification determined by the state of the world and the linguistic context shared by the speakers:

1. Readings in the domain of quantification are indexed by virtue of being present in the ICG via linguistic and non-linguistic clues that constitute shared information.
2. Additional readings are not normally indexed but are nevertheless decodable and inscribable according to the principles of semantics and logic that speakers share in the GCG. These readings, while derivable thanks to information that is part of the CG between participants, are not indexable in every communicative situation because they are not made relevant in the ICG.

The Immediate Common Ground is specific to a particular communicative situation; in this sense, the kind of information that is part of the ICG depends, case by case, on the linguistic and non-linguistic development of the conversation. The General Common Ground level is where we store information and knowledge that pertain to our use of language, in both explicit format, e.g. propositional knowledge, and implicit format, e.g. the heuristics and rules of logic we employ while speaking and communicating, along with conventions that we automatically follow in conversation, and information about the most common and frequent linguistic and non-linguistic conventions. The fact that implicit knowledge is part of the GCG is fundamental for our account of how Common Ground sheds light on the readings of plural covers. In the GCG we find the dual cover interpretation and the rules that can be used to derive other readings. This, obviously, does not imply that the formalization above is somehow part of the Common Ground among speakers; what is entailed in our competence with English, which is part of our knowledge in a broad sense, however, is the fact that we can make sense of a sentence like *The musicians wrote songs* with a double cover interpretation, according to which a writing event occurs, involving groups of musicians, and groups of songs as a result. It is at the level of ICG, on the other hand, that the specific readings are indexed. In other words, it is in specific communicative situation that specific cover readings can be understood.

With this distinction, we can succinctly describe the interpretation of sentences like (1) in different communicative situations, which also fits the empirical data.

(1)   The men wrote musicals.

In the first situation, Wyatt and Grace know about the careers of Rodgers, Hammerstein, and Hart, because they attend a class on History of Broadway together. While walking across campus, Wyatt brings up his favorite Broadway writers, Rodgers, Hammerstein, and Hart, and stars, Joan Roberts, Gertrude Lawrence, and Mitzi Green. In this situation, (1) is easily compatible with (39).

(39)   Hammerstein and Hart didn't write musicals individually or together.

The fact that Rodgers wrote musicals with Hammerstein and Hart respectively it is part of the GCG between Grace and Wyatt, and because of the conversation, this reading and the fact that they are the referents of *the men* is part of the ICG. The intersection of the available cover of (1) and the interpretation of (39) includes a true reading so the two sentences are compatible.

In the second situation, Wyatt is talking to Kara, who does not attend the class and is not well-versed in Broadway history, so she does not know who the three men and women are. When Wyatt brings up their names, and then utters (1), she will only get the cumulative reading by virtue of this reading being the interpretation of such a sentence in an ambiguous context as is specified in the GCG. If Wyatt uttered (39), the pair of sentences would not make sense: the intersection of the available cover readings of these sentences would be the empty set. To make sense of (1) and (39), Kara would need the correct cover reading to be made available, for example with the statement in (40).

(40)   Rodgers wrote musicals with Hammerstein, and he wrote musicals with Hart, but none of them wrote individually, and they never wrote all together.

With the distinction between ICG and GCG, we can also characterize the analyses of [10, 11, 24, 26] within a single framework. Gillon's [10, 11] assumption that both the beliefs and expectations of interlocutors and shape what readings are available corresponds to information in the GCG, while his assumption that context also shapes what readings are available corresponds to information in the ICG. Schwarzschild's [24] assumption that certain readings must be explicitly mentioned or otherwise salient because of non-linguistic discourse in order to be available also corresponds to our characterization of how the ICG shapes the domain of quantification. Finally, Sternfeld's [26] assumption that interlocutors choose particular readings in effort to make sentences true corresponds to the interaction of ICG and GCG, where interlocutors can invoke rules in the GCG to derive cover readings in the ICG. Lasersohn's [23] claim that certain readings are never available corresponds to readings that can be derived via rules in the GCG but are never brought into the ICG.

5   Conclusion

The proposed analysis provides a plausible explanation for why each condition was judged to be necessarily false in the empirical study. The ambiguous context in which the set-up sentences were presented was an empty ICG: There was no (non-)linguistic information that made any cover readings available, so the interpretation of the sentences evoked general linguistic principles in the GCG, which resulted in the default, double cover readings of the sentences in question. The follow-up sentences provided negative information about cover readings, however, because only the cumulative reading was available at this time, the intersection of the indexed cover readings in set-up sentences and the follow-up sentence was the empty set, and the follow-up sentences were judged to be false. What remains to be shown is whether or not certain intermediate readings are never judged to be true as Lasersohn [23] claims.

The fact that follow-up sentences with both *individually* and *together* were judged false significantly more frequently than those with only *individually* or *together*, is a phenomenon that must be accounted for. It might suggest that collective and distributive readings are more simple to derive than intermediate cover readings, which corresponds to the claim supported by many that these readings are more straightforwardly available—e.g. [10,20,21,24]. However, it seems that these readings might not be able to be assumed to be part of the interpretation in ambiguous contexts in light of the evidence found in this study, and therefore the following question remains open: Why are collective and distributive readings more simple to get than intermediate cover readings?

One possible explanation for the difference in judgments is the respective frequencies of overtly collective, distributive, and intermediate cover readings. Both the number of lexical modifiers that specify collective or distributive readings and their frequency of use lend to the intuition that these two minimal cover readings are more salient than intermediate covers. After all, it seems there are no lexical modifiers that index specific intermediate covers, and situations in which intermediate covers are salient are likely to be less frequent than situations in which collective or distributive interpretations are salient. A corpus study looking for the relative frequencies of these readings could validate this hypothesis.

References

1. Beck, S., Sauerland, U.: Cumulation is needed: a reply to winter (2000). Nat. Lang. Semant. **8**(4), 349–371 (2000)
2. Berio, L., Latrouite, A., Van Valin, R., Vosgerau, G.: Immediate and general common ground. In: Brézillon, P., Turner, R., Penco, C. (eds.) CONTEXT 2017. LNCS (LNAI), vol. 10257, pp. 633–646. Springer, Cham (2017). https://doi.org/10.1007/978-3-319-57837-8_51
3. Bianchi, C.: "Nobody loves me": quantification and context. Philos. Stud. Int. J. Philos. Anal. Tradit. **130**, 377–397 (2006)
4. Brennan, S.E., Clark, H.H.: Conceptual pacts and lexical choice in conversation. J. Exp. Psychol. Learn. Mem. Cogn. **22**(6), 1482–1493 (1996)
5. Clark, H.H., Schreuder, R., Buttrick, S.: Common ground and the understanding of demonstrative reference. J. Verbal Learn. Verbal Behav. **22**, 245–258 (1983)
6. Clark, H.H., Marshall, C.R.: Definite reference and mutual knowledge. In: Webber, B.L., Joshi, A.K., Sag, I.A. (eds.) Elements of Discourse Understanding. Cambridge University Press, Cambridge (1981)

Readings of Plurals and Common Ground   41

7. Gauker, C.: Domain of discourse. Mind **106**, 1–32 (1997)
8. Gerrig, R.J.: The scope of memory-based processing. Discourse Process. **39**, 225–242 (2005)
9. Gillon, B.S.: The readings of plural noun phrases in English. Linguist. Philos. **10**(2), 199–219 (1987)
10. Gillon, B.S.: Bare plurals as plural indefinite noun phrases. In: Kyburg, H.E., Loui, R.P., Carlson, G.N. (eds.) Knowledge Representation and Defeasible Reasoning, pp. 119–166. Springer, Dordrecht (1990). https://doi.org/10.1007/978-94-009-0553-5_6
11. Gillon, B.S.: Plural noun phrases and their readings: a reply to Lasersohn. Linguist. Philos. **13**(4), 477–485 (1990)
12. Grice, P.: Meaning. Philos. Rev. **66**, 377–388 (1957)
13. Grice, P.: Presupposition and conversational implicature. In: Cole, P. (ed.) Radical Pragmatics. Academic Press, New York (1981)
14. Horton, W.S., Gerrig, R.J.: Conversational common ground and memory processes in language production. Discourse Process. **40**(1), 1–35 (2005)
15. Horton, W.S., Gerrig, R.J.: Revisiting the memory-based processing approach to common ground. Top. Cogn. Sci. **8**, 780–795 (2016)
16. Kaplan, D.: Dthat. In: Uehling, T., French, P., Wettstein, H. (eds.) Contemporary Perspectives in the Philosophy of Language. University of Minnesota Press, Minneapolis (1978)
17. Kaplan, D.: Afterthoughts. In: Almog, J., Perry, J., Wettstein, H. (eds.) Themes from Kaplan, pp. 565–614. Oxford University Press, Oxford (1989)
18. Krifka, M., Musan, R.: Information structure: overview and linguistic issues'. In: Krifka, M., Musan, R. (eds.) The Expression of Information Structure, pp. 1–44. De Gruyter Mouton, Berlin (2012)
19. Krifka, M.: Nominalreferenz und zeitkonstitution. zur semantik von massentermen, pluraltermen und aspektklassen, universität münchen. Ph.D. thesis, Ph.D. dissertation (1986)
20. Landman, F.: Events and Plurality: The Jerusalem Lectures. Studies in Linguistics and Philosophy, vol. 76. Springer, Dordrecht (2000). https://doi.org/10.1007/978-94-011-4359-2
21. Lasersohn, P.: On the readings of plural noun phrases. Linguist. Inq. **20**(1), 130–134 (1989)
22. Lasersohn, P.: Plurality, Conjunction and Events, vol. 55. Springer, Dordrecht (1995). https://doi.org/10.1007/978-94-015-8581-1
23. Lasersohn, P. : Mass nouns and plurals. In: von Heusinger, K., Maienborn, C., Portner, P. (eds.) Semantics: An International Handbook of Natural Language Meaning, vol. 2, pp. 1131–1153. DeGruyter (2011). https://doi.org/10.1515/9783110255072.1131
24. Schwarzschild, R.: Pluralities, vol. 61. Springer, Dordrecht (1996). https://doi.org/10.1007/978-94-017-2704-4
25. Stalnaker, R.: Common ground. Linguist. Philos. **25**(5/6), 701–721 (2002)
26. Sternefeld, W.: Reciprocity and cumulative predication. Nat. Lang. Semant. **6**(3), 303–337 (1998)
27. Syrett, K., Musolino, J.: Collectivity, distributivity, and the interpretation of plural numerical expressions in child and adult language. Lang. Acquis. **20**(4), 259–291 (2013)
28. Tomasello, M.: Constructing a Language. Harvard University Press, Cambridge (2009)

Towards an Analysis
of the Agent-Oriented Manner Adverbial
sorgfältig ('carefully')

Ekaterina Gabrovska[(✉)]

Heinrich Heine University, Universitätstr. 1, 40225 Düsseldorf, Germany
egabrovska@phil.uni-duesseldorf.de

Abstract. The paper discusses the class of agent-oriented manner adverbials based on an investigation of the lexical meaning of the adverbial *sorgfältig* ('carefully'). The analysis proposes that the adverbial specifies the content of the action-plan of the agent and with this imposes restrictions on the manner of the modified event. The combination of intentionality and manner aspects in the meaning contribution of the modifier is explained based on Goldman's [9] theory of human action. The analysis is formalized in Frame Semantics as introduced in Löbner [14] and [15].

Keywords: Lexical semantics · Agent-orientation · Agent-oriented manner adverbials · Frame semantics

1 Introduction

The investigation of adverbial modifiers sheds light on the semantic representation of events, as well as that of agentivity. Especially interesting with respect to these two issues are agentive adverbials, such as subject-oriented and mental-attitude adverbials, which have been extensively explored in the literature (cf. Ernst [6], Geuder [11], Wyner [25] a.o.). The exact nature of the agent-relatedness in connection to manner adverbials, such as the class of German agent-oriented manner adverbials, e.g. *sorgfältig* ('carefully'), *geschickt* ('skillfully'), proposed in Schäfer [21], has been mostly left aside. These adverbials can provide us with new insights on the nature of manners of events as well as the role of the agent. What we aim for in this paper is to clarify the nature of agent-relatedness of agent-oriented manner adverbials as well as whether this agent-relatedness is connected to the manner of the modified event. We approach this issue by analyzing the meaning contribution of such adverbials based on a case study of the modifier *sorgfältig*.[1]

---

[1] Although *sorgfältig* is translated here primarily as *carefully*, there is no English adverb coinciding in meaning with the German modifier.

© Springer-Verlag GmbH Germany, part of Springer Nature 2019
J. Sikos and E. Pacuit (Eds.): ESSLLI 2018, LNCS 11667, pp. 42–61, 2019.
https://doi.org/10.1007/978-3-662-59620-3_3

1.1 Agent-Oriented Manner Adverbials

Schäfer [20,21] analyzes the class of agent-oriented manner adverbials (henceforth, AOMA), e.g. *sorgfältig* ('carefully') and *geschickt* ('skillfully')[2], as a subclass of manner adverbials, i.e. a kind of event modifier. The class of manner adverbials includes, in addition to the agent-oriented manner adverbials, also pure manner adverbials (Schäfer [21]).

(1) a. *Sie hat **laut** gesungen.*   [pure manner]
          she has loud sung
          'She sang loudly.'
     b. *Er löste   die Aufgabe **intelligent**.*   [AOMA]
          he solved the problem intelligent
          'He solved the problem intelligently.'
                                            (Schäfer [21]:49, ex. (1a, b))

As manner modifiers, the adverbials in (1) specify conceptual coordinates of complex cognitive structures (Schäfer [21]:51, cf. also Bartsch [3] and Geuder [11]). For example, the adverbial *laut* can specify the sound volume coordinate/aspect of an event concept (Schäfer [21]; cf. also Bartsch [3], also Morzycki [16], Piñón [22]). They can be questioned by a *Wie?*('How?')-question and do not scope over negation, as there has to be an event in order to modify its manner (cf. Pittner [23]:92, a.o.). Following Schäfer [21], sentences with manner adverbials can be paraphrased by the well-known *Wie-das-ist*-paraphrase (*The-way*-paraphrase) as in (2a) and *auf-ADJ-Art-und-Weise*-paraphrase (*in-ADJ-manner*-paraphrase) as in (2b):

(2) a. *Perta singt laut. = Wie Petra sings, das ist laut.*
          Petra sings loudly How Petra sings, that is loud
          'Petra sings loudly. = They way Perta sings is loud.'
                                            (Schäfer [21]:53, ex. (9))
     b. *Perta las   den Text sorgfältig. = Petra las   den Text auf eine*
          Petra read the text carefully   Petra read the text in a
          sorgfältige Art und Weise.
          careful   manner
          'Petra read the text carefully. = Petra read the text in a careful
          manner.'

The difference between AOMA and pure manner adverbials is that AOMAs require the highest ranked argument of the sentence to have control over the event (Schäfer [21]:59, cf. also Eckardt [5]). Hence, following Schäfer ([21]:61), the conceptual difference between pure manner and agent-oriented manner adverbials lies in the fact that the former are interpreted without "recourse to the properties of the agent".

---

[2] The two adverbials, *sorgfältig* and *geschickt*, are the cases, which are primarily used to discuss the class. In this paper, we concentrate on the meaning of *sorgfältig*, while *geschickt* is left for further work.

(3)   *Der Stein rollte schnell / *sorgfältig den Abhang hinunter.*[3]
        The stone rolled quickly / carefully   the hill   down
        'The stone rolled quickly / *carefully down the hill.'

(Schäfer [21]:59, ex. (26))

That AOMAs are agent-oriented is also indicated by the addition of an agentive *von*-phrase to the *Wie-das-ist*-paraphrase (Schäfer [21]:59). With pure manner adverbials this addition is not possible as shown in (4a).

(4)   a.   *Perta singt laut.* $\not\approx$ *Es war laut von Petra, wie sie singt.*
              Petra sings loudly It was loud of Perta, how she sings
              'Petra sings loudly. $\not\approx$ It was loud of Petra, how she sings.'

(Schäfer [21]:60, ex. (32))

        b.   *Perta las   den Text sorgfältig.* $\approx$ *Es war sorgfältg von Petra, wie*
              Petra read the text carefully   It was careful of Petra, how
              sie den Text las.
              she the text read
              'Petra read the text carefully. $\approx$ It was careful of Petra, how she read the text.'

In sum, AOMAs are manner adverbials with an orientation towards the agent of the event. As such, they have to be distinguished from subject-oriented (sometimes also called agent-oriented adverbials) and mental-attitude adverbials, such as *dummerweise* ('stupidly') and *widerwillig* ('reluctantly'), which are also considered to relate to the agent of the event (cf. Ernst [6], Schäfer [21], Geuder [11], Wyner [25] a.o.).

(5)   a.   *Dummerweise hat er geantwortet.*   [agent-oriented]
              Stupidly   has he answered
              'Stupidly, he answered.'
        b.   *Maria geht widerwillig schwimmen.*   [mental-attitude]
              Mary goes reluctantly swimming
              'Mary goes swimming reluctantly'

Following Schäfer [21] and Eckardt [5], subject-oriented adverbials are evaluative sentential adverbials which do not state anything about the manner of the event. Mental-attitude adverbials, on the other hand, are considered event adverbials which introduce the attitude of the agent towards the event (Ernst [6], Wyner [25]).

1.2   Analyses of AOMAs: State of the Art

Up to now, none of the representatives of the class of AOMAs have been analyzed in detail with respect to their lexical meaning. Even in English, adverbs like *carefully* and *skillfully* are discussed only in connection to the treatment of manner modifiers as atomic, unanalyzed predicates of events (cf. Eckardt [5] a.o.).

---

[3] The * signals unacceptability, whereas the # symbol means oddness due to meaning.

Given that *carefully* can have several translations in German, among which is *sorgfältig*, a complete parallel in the meaning of the two modifiers cannot be assumed. Hence, we take Schäfer [21] and Hansson [12] as a starting point instead of the literature concerned with manner adverbials in English. Both works are primarily concerned with a classification of German adverbial modifiers, but offer some assumptions with respect to the lexical meaning of *sorgfältig*. Both Schäfer and Hansson propose that (i) the modifier can describe the manner as well as the result of an event and (ii) that *sorgfältig* can attribute a property to the agent of the event.

As far as (i) is concerned, Schäfer [21], as well as Hansson ([12]), state that a clear separation between process and result modification seems impossible in connection with *sorgfältig*, at least in most cases. Schäfer [21] suggests that the modifier is a 'blend' between manner and result, adopting the notion 'blend' from Quirk et al. ([18]:560). He claims that the resultative passive construction, as in (6b), is entailed by the sentence as in (6a).

(6)   a.   *Peggy hat das Zelt sorgfältig aufgebaut.*
            Peggy has the tent carefully put up
            'Peggy put up the tent carefully.'
      b.   *Das Zelt ist sorgfältig aufgebaut.*
            The tent is carefully put up
            'The tent is put up carefully.'

We think that entailment is too strong as a relation between the sentences. Consider the following context: Peggy is careful in putting up the tent, but the latter is defect and collapses immediately after Peggy is done assembling it. In such a context the state of being carefully assembled does not hold of the tent even though Peggy was assembling it carefully.

Hansson [12] suggests that the result component is available only with potentially resultative verbs as *aufstellen* ('to arrange') in (7a) (*aufbauen* in (6) is also resultative). Such verbs are analyzed in Hansson [12] as having a process (p) and a result (r) phase part such that p, r $\subseteq$ e. Whether *sorgfältig* modifies the process or the result phase part is determined by the context, hence the modifier has to apply to either the process or result phase but not both at the same time. When *sorgfältig* modifies verbs like *arbeiten* ('to work'), which provides just a process component (cf. Hansson [12] for *lesen* ('to read') and *malen* ('to paint')), however, at least some inference concerning a possible result is involved.[4]

(7)   a.   *Sie stellte die Stühle sorgfältig auf.*
            She put   the chairs carefully   up
            'She arranged the chairs carefully.'
                                   (Hansson [12]:133, ex. (22))
      b.   *John hat sorgfältig gearbeitet.*
            John has carefully worked
            'John worked carefully.'

---

[4] Example (7b) can have two different interpretations: (i) the speaker is understood as having seen John working *sorgfältig* and therefore expects a good result, while (ii) it is seen in the result of the work that John has worked *sorgfältig*.

Given that there is no entailment regarding the result in (6) and that "non-resultative" verbs, like *arbeiten* ('to work'), as in (7b), also invite the assumption of a result, we propose that *sorgältig* is evoking an inference in connection to some outcome as a result of the performed action as we show in more detail in Sect. 3.

Considering (ii), both authors observe that the modifier not only seems to state something about the agent, but that inanimate objects are usually not able to do things. Consider again example (3): a stone cannot be interpreted as being intentional and having control over the rolling, except if personified, and therefore cannot act *sorgfältig* (cf. Schäfer [21] and Hansson [12]). A normal, i.e. non-agentive, manner modifier, however, is fine as it does not refer to properties of the agent as illustrated with *schnell* in (3).[5]

Given (i) and (ii), Schäfer's [21] and Hansson's [12] suggestions that the modifier is to be treated as a predicate over manners or over the process part of an event, respectively, seem inadequate. Neither Schäfer's [21] or Hansson's [12] analysis, nor the standard analysis of manner adverbs as predicates of events (cf. Eckardt [5] a.o.), accounts for the observations that *sorgfältig* is both agent-oriented and modifies the event, or the observation that the modifier invites an inference with respect to the result of the event.

1.3   Proposal

We propose that there is an intentionality component as part of the lexical meaning of the modifier which has not been recognized by the literature so far. This component captures the observation that *sorgfältig* can modify only events involving an agent. We suggest treating the modifier as building on and enriching the meaning of the adverbial *absichtlich* ('intentionally') with a manner meaning component. As such, *sorgfältig* ('carefully') demands the participation of an intentional and controlling agent in the event and specifies the interpretation of potentially intentional verbs[6] or applies to ones which can only be interpreted as intentional. Potentially intentional verbs are verbs which do not lexicalize intentionality/agency, e.g. *kill*, whereas verbs like *murder* are seen as lexically intentional (cf. Van Valin [24], Buscher [4]).

---

[5] Hansson [12] questions the agent-orientedness of the modifier based on the following example:

(i)   *Die Autowaschanlage wusch   das Auto sorgfältig.*
       The car wash   washed the car   carefully
       'The car wash cleaned the car carefully.'

(Hansson [12]:136, ex. (46))

Hansson [12] suggests that in such cases the modifier is used metaphorically. How such examples should be analyzed is not discussed in detail in this work, but a treatment that does not pose a problem for the view that *sorgfältig* is agent-oriented is proposed in Gabrovska [8].

[6] The assumption is also made for mental-attitude adverbials (cf. Buscher [4]).

We see *sorgfältig* as imposing certain restrictions on the agent's goal and the way the latter is to be achieved. We propose that these restrictions are best captured by Goldman's [9] 'action-plans' which are at the heart of his analysis of intentional action. The reference to action-plans is considered to be supplied by the agent-orientation of the modifier. A manner specification is achieved by the restrictions imposed on the way the goal is to be realized. This treatment allows us to capture all meaning components of *sorgfältig*, i.e. the intentionality and manner components as well as the result inference.

We proceed with a brief overview of Goldman's [9] ideas and their reinterpretation by Löbner [15]. With these prerequisites, we explain how they can be used for an analysis of the lexical meaning of *sorgfältig* in Sect. 3. The proposal is formalized in the framework of frame semantics in Sect. 4. The formalization we propose is schematic, in that we only describe the final effect of modification by *sorgfälltig* on the event representation, the additions to the event representation that are triggered by use of the adverbial.

2   A Theory of Human Action

2.1   Goldman's [9] view on actions

Goldman [9,10] argues that the example in (8) involves different acts done by the same agent at the same time.

(8)   John wakes up Mary
                     ↑
            John turns on the light
                     ↑
            John flips the switch
                     ↑
            John moves his hand

                    (Example partly adopted from Goldman [9])

The upward arrow in the example stands for an intuitive relation between John's acts that Goldman [9] calls 'level-generation'. Level-generation captures that intuition that (in the appropriate circumstances), by moving his hand, John flips the light switch, and thereby turns on the light and so on. Goldman suggests that this 'by' locution is diagnostic of a level-generation relation between acts. Level-generation is a transitive, asymmetric and irreflexive relation and gives rise to hierarchical structures, called 'act-trees'. The act of moving one's hand is considered by Goldman [9] a basic act, which itself is not generated by any other act. Basic acts are at the bottom of every act-tree.

Goldman's [9] theory offers a mechanism by which intentional (as well as unintentional) acts can be explained. The mechanism is referred to by the notion 'action-plan'. An action-plan consists of an 'action-want', i.e. the want to realize a certain act, and the 'level-generational beliefs' of the agent.

The level-generational beliefs can be also represented in a hierarchical structure as in (8), called 'projected act-tree' in Goldman ([9]:56). We understand projected act-trees as mental representations. They represent hypothetical acts which are supposed to be performed if the agent is to perform some certain basic act (Goldman [9]:56). The projected act-tree is constructed based on the level-generational beliefs of the agent. Considering again example (8), if John acts intentionally then he would have an action-plan, i.e. a mental representation of his action, where based on his level-generational knowledge, he believes that his moving his hand would generate his flipping of the switch, which would in turn generate his turning on the light and so on.

Following Goldman [9], if John's action-plan matches his actual action, then the latter is intentional. If, however, there is a mismatch between plan and action execution, the acts which differ from the acts in the plan and all acts they generate are not intentional. Let us illustrate this with an example: assume John executes the action in example (8) and has the following plan *flip the switch* ↑ *turn on the cooling fan*. First, if John is to turn on the cooling fan intentionally, then he has to have the want to turn it on. This want is part of John's action-plan and causes his actual action (see Goldman [9]). We consider this want to realize a certain act as having that act as a target. The latter is represented as part of the projected act-tree (the topmost act in the tree).

John can further have different options for turning on the cooling fan, but since he decides to use the switch, he believes that this method is going to realize the act he desires, i.e. the target of his action-want. Given this plan, John proceeds to act and flips the switch, but instead of generating the act of turning on the cooling fan, his flipping of the switch generates the act of his turning on of the light which itself constitutes his waking up Mary. With this John's actual acts does not match his plan on all levels besides the first one. Hence, only the flipping of the switch can be considered intentional, whereas all other acts of John are unintentional.

This treatment of intentional action enables us to assume that adverbials, like *absichtlich* ('intentionally') or *versehentlich* ('accidentally, inadvertently'), refer to the action-plan of the agent and at the same time state that the plan matches (for *absichtlich*) or does not match (for *versehentlich*) the actual action of the agent, respectively (see Gabrovska and Geuder [7] for a detailed analysis of *absichtlich*). This treatment also captures the difference between an action and a happening, e.g. coughing due to sickness (happening) and coughing as a sign to somebody (act). Only actions involve an action-plan in their representation. Hence, *absichtlich/versehentlich husten* ('coughing intentionally/inadvertently') is interpreted as an action as both modifiers are assumed to refer to plans. If a coughing is interpreted as a happening both modifiers cannot apply as things that happen to one are not seen as intentional or unintentional (they are not under the control of the agent and the agent does not/cannot fail to control them).

2.2   Löbner's [15] Reinterpretation of Goldman [9]

Goldman's [9] theory of human action is originally a contribution to the philosophical discussion on act individuation, but as stated in Löbner ([15]:2) it can be reinterpreted as a cognitive theory, i.e. "a theory of the cognitive categorization of action."

Consider again the example in (8). John's hand movement is what is observable in the real world and, given the appropriate circumstances, it can be categorized as an act of flipping the switch, turning on the light, and waking up Mary. Under other circumstances this movement might be categorized as something else. In our example only this one movement of John, categorized in different ways, i.e. as acts of different types (henceforth, act-TTs as called in Löbner [15][7]), is observable. This is one doing that "constitutes a combination of distinct act-tokens of distinct act-types" (Löbner [15]:4).

Goldman's [9] level-generation is considered by Löbner [15] to be a cognitive process, i.e. a process of relating categorizations for an act-token, from which the conceptual relation of 'c-constitution' ('c' standing for circumstances) results. This relation holds between act-TTs related by level-generation. C-constitution states that under the given circumstances an act-TT a/A c-constitutes an act-TT b/B iff by doing a/A the agent exemplifies act b of type B (Löbner [15]:12).[8] The relation is transitive, asymmetric and irreflexive and gives rise to hierarchical structures as in example (8), called cascades in Löbner [15].

3   Analyzing *sorgfältig*

So far we have suggested that the modifier *sorgfältig* has manner and intentionality meaning components, as well as that it evokes an inference concerning the agent's goal. Given Goldman's [9] theory and Löbner's [15] reinterpretation of it, we assume that the intentionality component can be captured via action-plans. The manner component is considered to be dependent on the target of the agent's action-want, i.e. the agent's goal. We turn to these issues in the following section.

3.1   Agent-Orientation of the Modifier

In Sect. 1, we stated that in the case of AOMAs the highest-ranked argument must be in control of the action (cf. Schäfer [21]). However, prior literature has not examined in detail the properties of this argument and whether it behaves like a prototypical agent. Mental-attitude adverbials, for example, were long assumed to require an agent, but as shown by Buscher [4], this is not the case.

---

[7] According to Löbner [15]:4 TTs (tokens-of-a-type) are "ordered pairs of an entity and a type such that the entity is of this type." Löbner [15] sees the assumption of act-TTs as very natural: VPs provide descriptions of events by giving their types.

[8] Adopting the notation in Löbner [15], in the pair a/A a is an act-token, whereas A is stands for the act-type.

Following her analysis, the subgroup of assimilative adverbials, like *widerwillig* ('reluctantly'), demand a controlling participant, whereas intentional adverbials, like *absichtlich* ('intentionally'), a participant initiating the event, but none of them demand an agent.[9] Thus, it has to be clarified whether an agent (an intentional, volitional, and sentient being[10]) is involved in modification with *sorgfältig*. Consider the following examples inspired by Buscher [4]:

(9)   a. #*John hat sich   sorgfältig verlaufen.*
              John has himself carefully get-lost
              'John got lost carefully.'
          b. #*John rutschte sorgfältig aus.*
              John slipped   carefully PART
              'John slipped carefully.'

The examples in (9a) and (9b) show that *sorgfältig* cannot modify verbs denying control, as well as unaccusative verbs which are not agentive (cf. Buscher [4]:103ff a.o.), respectively. Example (9) speaks in favor of the view that verbs modified by *sorgfältig* ('carefully') have to allow for an intentional interpretation (or have only an intentional interpretation, considering verbs like *arbeiten* ('to work') or *planen* ('to plan')).

(10)   a. *Das Kind hat das Bild   absichtlich/unabsichtlich*
              The child has the picture intentionally/unintentionally
              zerschnitten.
              cut
              'The child cut the picture intentionally/unintentionally.'
          b. *Das Kind hat das Bild   sorgfältig zerschnitten.*
              The child has the picture carefully cut
              'The child cut the picture carefully.'

Example (10a) shows that *zerschneiden* ('to cut') can be done intentionally or unintentionally. However, when the verb is modified by *sorgfältig*, (10b), only an intentional interpretation is possible. Consider the following context where a child holds two pictures, one of his mother and one of his father, but is not aware that she holds them both as the picture of the mother is on top of the picture of the father hiding it completely. Now, if the child cuts *sorgfältig* the picture of her mother, it does not follow that she cuts *sorgfältig* the picture of her father in the same manner. Likewise, if the child cuts the picture of her mother intentionally, this does not mean that she also cuts the picture of her father intentionally. On the contrary, in the given context, the cutting of her father's picture is unintentional and hence it cannot be *sorgfältig*.

Altogether, the data presented so far is seen as supporting the view that an agent has to be present whenever the modifier *sorgfältig* ('carefully') is used. The examples in (10) indicates that when combined with verbs which can be

---

[9] See Buscher [4] for a discussion and supporting data.
[10] We follow Van Valin and Wilkins [24] as far as these notions are concerned.

interpreted as either intentional or unintentional, the modifier specifies the interpretation to the intentional variant.[11] This means that the agent is acting intentionally. We use Goldman's [9] action-plans to capture this observation. Hence, whenever an agent is acting intentionally, she has an action-plan. This means that the agent has an action-want and chooses with respect to her beliefs the appropriate way of achieving her goal.

3.2 The Inference About the Agent's Goal

The agent, if acting *sorgfältig* ('carefully'), has to meet further requirements via her action-want and the way her beliefs realize that action-want.

(11)   a.   *Wir haben sorgfältig gearbeitet, weil   wir gute Ergebnisse*
           We   have   carefully   worked,   because we   good   results
           erzielen wollten.
           achieve   wanted
           'We worked carefully, because we wanted to achieve good results.'

       b.   # *Wir haben sorgfältig gearbeitet, weil   wir schlechte Ergebnisse*
           We   have   carefully   worked,   because we   bad   results
           erzielen wollten.
           achieve   wanted
           'We worked carefully, because we wanted to achieve bad results.'

Both examples verbalize the agent's desire: achieving good/bad results. The difference in acceptability between (11a) and (11b) is due to the quality of the result. Under normal circumstances a *sorgfältig* ('careful') action is unlikely to be performed if the quality of the result is intended to be low. Assuming a context where the agent actually desires to achieve bad results because, for example, she wants to prove some certain theory wrong, the sentence in (11b) can be felicitous. The observation is further supported by the following data:

(12)   a.   *Wir haben sorgfältig gearbeitet, daher   haben wir gute*
           We   have   carefully   worked,   therefore have   we   good
           Ergebnisse.
           results
           'We worked carefully, therefore our results are good.'

       b.   # *Wir haben sorgfältig gearbeitet, daher   haben wir schlechte*
           We   have   carefully   worked,   therefore have   we   bad
           Ergebnisse.
           results
           'We worked carefully, therefore our results are bad.'

---

[11] As noted above, Buscher [4] states that mental-attitude adverbials, like *absichtlich* ('intentionally'), also specify the interpretation of verbs. We assume that this common feature between the two classes can be explained by the use of the same mechanisms (see Gabrovska and Geuder [7]).

Comparing (12a) and (12b) we see that it is the quality of the achieved results which indicates the oddness of the latter. Data like this shows that a *sorgfältig* action is expected to lead to a "good" result.[12] Although the examples in (12b) and in (11b), are not completely uninterpretable, additional context is necessary to make them acceptable. (12a) and (11a) are fine on their own.

This specification of the result as one with a high quality is, however, a context-dependent inference as indicated by the next example as well as the examples discussed in Sect. 1.2.

(13)   *John hat sein Zimmer sorgfältig geputzt, trotzdem   hingen*
         John has his   room   carefully cleaned nevertheless hung
         Spinnweben an der Decke.
         spider webs on the ceiling
         'John cleaned his room carefully but nevertheless spider webs hung from the ceiling.'

Although the cleaning is said to be *sorgfältig*, the achieved result is not seen as "good" enough. Such data indicates that the target of the agent's want does not have to be realized. Nevertheless, the cleaning itself is still intentional and its method of implementation still suitable for the achieving of at least part of the relevant aspects of the intended result. Furthermore, native speakers, when confronted with such data, tend to assume reasons outside of the control of the agent as an excuse. Hence, in the example above, the agent is assumed to be unable to clean the spider webs due to the height of the ceiling, for example. This leads to the conclusion that the quality of the result is measured with respect to the context, the capabilities of the agent, their knowledge, and instruments available to carry out the action.

3.3 The Manner Component

The inference of a "good" result evoked by *sorgfältig* ('carefully') also depends on the manner in which the action is realized. We call this manner component the method of realization of the desired act and follow Sæbø [19] in defining method as steps leading to the accomplishment of an activity, i.e. a procedure.

Taking into account Goldman's [9] theory, we stated in Sect. 2 that the relation between acts is phrased by the use of 'by' (translated as *indem* in German). In the example *John turns on the light by flipping the switch by moving his hand*, 'by' is assumed to state that the act of flipping the switch is the method of realizing the act of turning on the light (cf. Goldman [9]). This means that the act of flipping the switch is suitable for generating the act of turning on the light under the given circumstances. Similarly, all other abstract actions are underspecified with respect to their methods of realization (cf. Löbner [15]).

---

[12] It has to be noted that *gute Ergebnisse* ('good results') or *having good results* is not an act as required by Goldman ([9]: 52f); the corresponding desired act is *to achieve/gain good results*.

(14)   a.   *Die Rüben   werden sorgfältig gereinigt, indem man sie   einige*
           The turnips shall   carefully cleaned,   by   one   them several
           Minuten lang im   Wasser läßt,   dann wäscht und
           minutes long in.DAT water   leaves, then   washes and
           abbürstet.[13]
           brushes down
           'The turnips are cleaned carefully, by leaving them in water for
           several minutes and then washing and brushing them down.'

   b.   #*Die Rüben   werden sorgfältig gereinigt, indem man sie   nur mal*
         The turnips shall   carefully cleaned,   by   man them just
         kurz unters Wasser hält.[14]
         short under water   hold
         'The turnips are cleaned carefully, by holding them under the tap
         just a little bit.'

Example (14a) supplies the method suitable of realizing a *sorgfältig* cleaning.
The method is complex and consists of the sequential realization of three acts:
putting in water, washing, and brushing down. However, not all methods implementing a cleaning are available when an action is said to be done *sorgfältig*
('carefully'). This is not unexpected, considering that the result of the action
should have a rather high quality. It is only natural that if a "good" result
is expected, not every method can guarantee high quality, e.g. (14b). Nevertheless, the number of possible methods is usually not reduced to exactly one
by the adverbial. Rather, the set of methods implementing a certain action is
specified more concretely and therefore reduced in the number of elements compared to the set containing all possible methods. Here again, the method and
its suitability depend on the context in which the action is performed a.o. (cf.
Sect. 3.2). Existing (social) conventions and rules may also have an impact on
the suitability.

4   A Frame Analysis

4.1   The Formalization

The proposal, as presented so far, is formalized in the framework of Düsseldorf
Frame Semantics, which adopts and develops Barsalou's [2] work on the representation of knowledge in human cognition. The theory is based on the Frame
Hypothesis, which states that frames are the general format of representation in
human cognition (Löbner [14]).

---

[13] https://books.google.de/books?id=FXGXBwAAQBAJ&pg=PA133&lpg=PA133&
    dq=%22sorgf%C3%A4ltig+gereinigt,+indem%22&source=bl&ots=e_6WSP7pZ1&
    sig=JT1a1IywRLv4QFJTUqW7eapOwRU&hl=de&sa=X&ved=0ahUKEwjo-
    KPH5o7ZAhXD1qQKHd3ACwQQ6AEIJzAA, last accessed 27.06.18.

[14] Thanks to Wilhelm Geuder for the example.

Frames. In this framework, frames are recursive attribute-value structures (Petersen [17], Löbner [14]). They can be represented as diagrams or attribute-value matrices.[15] The manner of representation adopted here are diagrams, hence a frame is represented as a network of nodes connected by labeled arcs. The nodes stand for values, whereas arcs stand for attributes.

Attributes in frames are functions from individuals to individuals (cf. Löbner [15]). An example for an attribute in a verbal frame is AGENT. The agent thematic role is known in semantics to be a function from events to individuals, hence it naturally qualifies as an attribute (cf. Gabrovska and Geuder [7]).

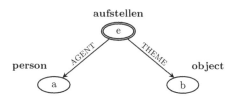

Fig. 1. Simplified partial frame representation for *aufstellen* ('to arrange')

The frame in Fig. 1 represents an event. The central node of the frame, marked by the double ellipse, has the value 'e' which stands for the concrete *aufstellen*-event. The node carries a type specification **aufstellen**, which states that the event 'e' belongs to the subset of the universe referred to by the meta-language predicate **aufstellen** (cf. also Anderson and Löbner [1]).

The event has two arguments, which are represented by the attributes AGENT and THEME. Both attributes have values, 'a' and 'b', which stand for the respective individuals. The agent of the event is of type **person**, whereas the theme is of type **object**. As frames are recursive structures, the latter two nodes themselves can also be represented via frames and have attributes of their own.

The frame in our example is seen as a description of a concrete individual, but frames have a double nature as stated in Löbner [15]. Apart from being read as a description of an individual (the token reading), they can be read as a description of a type (the type reading). On the type reading, frames describe a potential token of that type and with this they give a description of the type. Löbner's [15] act-TTs, as introduced in Sect. 2, have the same double nature.

Cascades as Frames. Löbner's [15] cascades are represented in frame theory as tree-like structure of frames which are related to each other via c-constitution. As it is whole frames that are related, cascades are second-order frame structures. Each of the frames in a cascade describes an act-TT: the values of the central nodes of the frames ('a_0', 'a_1', 'a_2', 'a_3') stand for the acts, whereas the whole frames describe the types of that acts. Consider our example from the beginning of Sect. 2 represented as a cascade in Fig. 2.

---

[15] Frames can also be translated into first-order predicate logic formulas. For one approach, see Löbner [14].

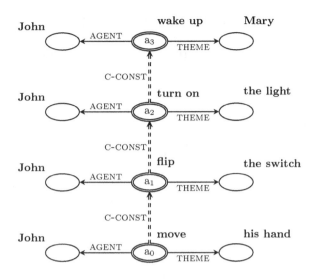

Fig. 2. A simplified cascade representation of example (8)

The cascade in Fig. 2 represents that under given circumstances (e.g. the switch is not out of order), the act of *moving his hand* by John constitutes the act of *flipping the switch* by the same agent. Likewise, the latter act under circumstances (e.g. there is a bulb in the lamp) constitutes the act of *turning on the light*, which constitutes the act of *waking up Mary*, if, for example, Mary is sensitive to light. The c-constitution relation is marked in the diagram by dashed double arrows (we use double dashed arrows for second-order relations and single arrows for frame attributes) pointing from the generating act to the generated one. As the relation holds between act-TTs the whole frames are related, although this relation is represented by a connection between their central nodes. The act-TTs a_1/'John flips the switch', a_2/'John turns on the light' and a_3/'John wakes up Mary' are each represented by a very simplified first-order frame.

The cascade approach proposed by Löbner [15] can be also applied to verb semantics as discussed by the author himself. Following this discussion, the lexical meaning of most, if not all, action verbs should be represented as cascades. This is due to the abstractness of the (non-basic) act-types they denote. Such non-basic act-types require that another act of some other type generates them (Löbner [15]). Basic acts, on the other hand, are not implemented by other acts and therefore do not involve c-constitution. These suggestions can be intuitively illustrated on the meaning of the verb *wake up* as in *John woke up Mary*: a waking up of somebody always involves some other action like a gentle shaking, for example, in order to come about. The shaking itself also has to be realized by another act, like moving one's hand. With this we arrive at the level of bodily movements. These can be considered basic as humans do not seem to conceptualize below that level[16]. The lexical meaning of *wake up*, leaves the type of the generating act(s) unspecified, but includes the information that there would have to be a generating level. Hence, the representation in Fig. 2 is not the lexical meaning of *wake up*, as the specification of the generating levels comes from the context during the interpretation.

4.2 A Model for *sorgfältig*

Modification with *sorgfältig* indicates that the modified event is to be interpreted as an intentional action (cf. Sect. 3). We adopt action-plans as proposed by Goldman [9] for the analysis of intentional action, which means that the representation of an action has to involve a plan component. In contrast, events which are not to be classified as actions do not involve an intentionality component, i.e. there is no plan in their representation.

The frame in Fig. 3 has a central node typed ***sorgfältig* action**. The possible values of the central node are characterized by the two attributes PLAN and EXECUTION. The PLAN attribute introduces the action-plan, more precisely the level-generational beliefs construction including the target of the agent's want (the goal of the agent), while the values of the EXECUTION attribute represent the 'physical' realization of the action. The action as well as the action-plan are represented as cascades. Thus, at the present stage of the analysis, the values in the plan and the execution nodes are cascades, cf. Fig. 5.

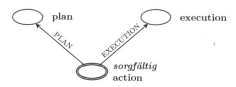

Fig. 3. A *sorgfältig action* frame skeleton

Following Goldman ([9]:59), an action is intentional if realized as conceived in the agent's action-plan. With respect to our model, this means that the plan and action representation have to match. This is achieved by the use of the comparator $©_{=t}$, which checks whether the acts at the respective levels have the same type or not.

Comparators are defined by Löbner [14] as special (first-order) two-place attributes with arguments of the same sort. They return comparison values, such as '=', '>', and '<' (not to be confused with the relations that these symbols are conventionally taken to denote, cf. Löbner [14]).

As the comparator $©_{=t}$ is comparing frames (the type is described by the whole frame), it is a second-order comparator returning the values '=' or '≠'.

---

[16] This assumption might be debatable, however we refrain from discussion of this topic here as this would lead us away from the goal set for the paper.

Whenever two acts match in type the comparator returns the value '=', otherwise '≠'. The comparator is represented in frames as in Fig. 4.

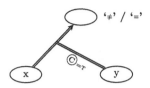

Fig. 4. The comparator ©$_{=t}$

As discussed in Sect. 3.2, the result does not have to be actually achieved, but only desired. This means that not all levels of the cascade need to match in type: if the result is not realized (or there is no information about its realization), there is no act higher than the modified one in the execution cascade. What *sorgfältig* adds is that there has to be a goal in the plan representation and that the method used to achieve that goal is suitable and is realized as conceived in the plan.

Besides indicating that the modified event is to be interpreted as an intentional action, *sorgfältig* imposes restrictions on the target of the agent's want and the method of realization of that target. The values of the highest node of the action-plan cascade (the node of type **achieve R$^+$**) are constrained to high quality results, hence the node representing the goal of the agent is typed as **achieve R$^+$**. Such results cannot be intentionally achieved by doing just anything, hence the restriction on the result values leads to a restrictions of the method values. That the method values are restricted is indicated by the type label **suitable method**.

Figure 5 shows a schematic representation of a *sorgfältig* action. The goal in the plan cascade as well as the method in the plan and in the execution cascade are unspecified. Their exact content is provided by the context during interpretation.

That the agent is acting intentionally is represented by the *act* and *method* frames on the right hand side being congruent with (part of) the plan on the left. As the method of goal implementation matches the respective level of the realization, the restrictions imposed by the modifier are realized in the agent's action.

The result component, if realized, has to be generated by a node of type **act** and has to match the intended result. For the sake of completeness, we use dotted lines in the frame to mark the possibility that there might be a result. If the context supplies information about the result, the latter is represented as usual, i.e. normal (or dashed) lines instead of dotted lines. If there is no information, there is no need to represent it.

The restrictions the modifier imposes on the values of the result are additionally represented by the gray boxes in Fig. 5. The frame in Fig. 5 is a representation for an action that includes all additions triggered by the modification with *sorgfälltig*.

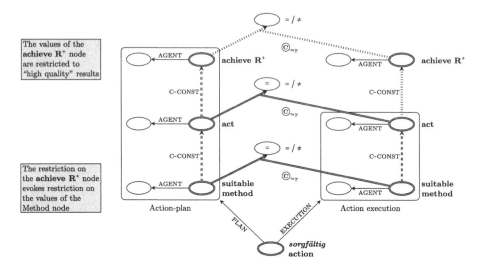

Fig. 5. An action with a action-plan

To illustrate the analysis, we use an example where the method is supplied via an *indem*-phrase. The frame in Fig. 6 partly represents the meaning of example (15). The THEME and other attributes are omitted for readability.

(15)   *John hat   sein Zimmer sorgfältig geputzt, indem er gesaugt   und*
       John have his   room   carefully cleaned by   he vacuumed and
       alle Spinnweben entfernt hat.
       all   spider webs removed had
       'John cleaned his room carefully by vacuuming and removing all spider webs.'

Sorgfältig indicates that John's cleaning is an action, hence the frame representation of the cleaning has to involve a plan. John (⚲) in our example is the agent, who is cleaning *sorgfältig* according to his own criteria and knowledge of the relevant conventions and rules. The method of realization, provided by the *indem*-phrase, is complex and consists of *saugen* ('to vacuum') and *Spinnweben entfernen* ('to remove spider webs'). The method is represented here as one frame for simplicity, but *vacuum* and *remove* each would have to be represented by frames in a fully specified analysis. It is still an open issue how complex methods should be best represented in frames.

John's goal (the result that is to be achieved) is specified in so far as it would involve something like having a nicely cleaned room instead of a superficially cleaned room (the dirt is vacuumed away instead of swept under the carpet, all the spider webs were removed instead of only the easily visible ones, etc.), due to modification by *sorgfältig*.

The lowest level in Fig. 6 stands for the method, i.e. vacuuming and removing spider webs. Due to the restriction of the agent's goal, the availability of possible

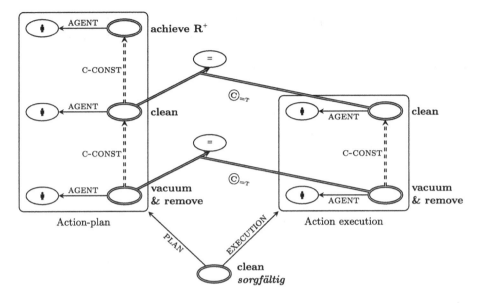

Fig. 6. Connecting action and action-plan

methods is also restricted. Available are only methods which can actually generate the goal. In our case, *saugen* ('to vacuum') and *Spinnweben entfernen* ('to remove spider webs') belongs to this set of methods, while sweeping the dirt under the carpet and removing the easily visible spider webs do not.

The intermediate level in both cascades in Fig. 5 represents the cleaning, which is implemented by vacuuming and removing spider webs. On top of the action-plan cascade, generated by the cleaning, is the result *achieve having a clean room.*The action is interpreted as being intentional because the plan and execution match; the comparator returns the value '=' for the relevant levels.

5   Summary and Outlook

So far we have proposed that intentional agents have action-plans as proposed by Goldman [9] and that *sorgfältig* ('carefully') restricts and relates the desire and the belief components of the action-plan in a specific way, namely that the method chosen for the realization of the desire has to be suitable with respect to the intended quality of the result. Hence, due to its intentionality component, *sorgfältig* refers to the plan of the agent and at the same time, due to its manner component, it specifies how the action should be realized with respect to the target of the action-want. The tentative formalization of the proposal is realized in the framework of frame semantics as in Petersen [17] and Löbner [14,15] and is capable of capturing agent-relatedness as well as its impact on the manner of events.

The tentative analysis presented here still leaves a number of open questions for future work, but it has the potential of explaining the meaning contribution of not only *sorgfältig* ('carefully'). We assume that agent-oriented manner adverbials refer to the plan of the participating agent and relate the target of the action-want as part of that plan to the method of realization. The difference between representatives of this class is then the way they restrict the agent's goal and the impact of this restriction on the method of realization. Furthermore, the meaning contribution of mental-attitude adverbials, like *absichtlich* ('intentionally'), can also be captured by using action-plans as shown by Gabrovska and Geuder [7].

Acknowledgements. Many thanks to Sebastian Löbner, Wilhelm Geuder, Curt Anderson, the anonymous reviewer as well as all my informants for the long discussions, the comments, the data, and all other kinds of help. The work is supported by DFG CRC 991 "The Structure of Representations in Language, Cognition, and Science," project B09.

References

1. Anderson, C., Löbner, S.: Roles and the compositional semantics of role-denoting relational adjectives. In: Sauerland, U., Solt, S. (eds.) Proceedings of Sinn and Bedeutung, vol. 22, pp. 91–108 (2018)
2. Barsalou, L.W.: Frames, concepts, and conceptual fields. In: Lehrer, A., Kittay, E.F. (eds.) Frames, Fields, and Contrasts: New Essays in Semantic and Lexical Organization, pp. 21–74. Lawrence Erlbaum Associates, Hillsdale (1992)
3. Bartsch, R.: The Grammar of Adverbials. North Holland Linguistic Series 16. North-Holland, Amsterdam, New York, Oxford (1976)
4. Buscher, F.: Kompositionalität und ihre Freiräume: Zur flexiblen Interpretation von Einstellungsadverbialen. Doctoral dissertation, Universität Tübingen (2016)
5. Eckardt, R.: Event semantics. Linguistische Berichte, Sonderheft, vol. 10, pp. 91–128 (2002)
6. Ernst, T.: The Syntax of Adjuncts. Cambridge University Press, Cambridge (2002)
7. Gabrovska, E., Geuder, W.: Andverbs of intentionallity, submitted
8. Gabrovska, E.: A frame-based analysis of agent-oriented manner adverbials in German. Doctoral dissertation, Heinrich-Heine Universität Düsseldorf, in preparation
9. Goldman, A.I.: A Theory of Human Action. Prentice-Hall INC., New Jersey (1970)
10. Goldman, A.I.: Action, causation, and unity. Noûs **13**, 261–270 (1979)
11. Geuder, W.: Oriented Adverbs. Issues in the Lexical Semantics of Event Adverbs. Doctoral dissertation, Universität Tübingen (2002)
12. Hansson, K.: Adverbiale der Art und Weise im Deutschen. Eine semantische und konzeptuelle studie Göteborger. Germanistische Forschungen 48. Göteborgs Universitet, Acta Universitatis Gothoburgensis (2007)
13. Löbner, S.: Evidence for frames from human language. In: Gamerschlag, T., Gerland, D., Osswald, R., Petersen, W. (eds.) Frames and Concept Types. Studies in Linguistics and Philosophy, vol. 94, pp. 23–67. Springer, Heidelberg, New York (2014). https://doi.org/10.1007/978-3-319-01541-5_2

14. Löbner, S.: Frame theory with first-order comparators: modeling the lexical meaning of punctual verbs of change with frames. In: Hansen, H.H., Murray, S.E., Sadrzadeh, M., Zeevat, H. (eds.) TbiLLC 2015. LNCS, vol. 10148, pp. 98–117. Springer, Heidelberg (2017). https://doi.org/10.1007/978-3-662-54332-0_7

15. Löbner, S.: Cascades. Goldman's level-generation, multilevel conceptualization of action, and verb semantics. In: Löbner, S., Gamerschlag, T., Kalenscher, T., Schrenk, M. Zeevat, H. (eds.) Cognitive Structures - Linguistic, Philosophical, and Psychological Perspectives, Springer, Dordrecht. Accepted by the volume editors

16. Morzycki, M.: Modification. Cambridge University Press, Cambridge (2016)

17. Petersen, W.: Representation of concepts as frames. In: Gamerschlag, T., Gerland, D., Osswald, R., Petersen, W. (eds.) Meaning, Frames, and Conceptual Representation. Studies in Language and Cognition. Düsseldorf University Press, Düsseldorf (2015). (Commented reprint of Petersen, W.: Representation of concepts as frames. In: Skilters, J., Toccafondi, F., Stemberger, G. (eds.) Complex Cognition and Qualitative Science. The Baltic International Yearbook of Cognition, Logic and Communication, vol. 2, pp. 151–170. University of Latvia (2007))

18. Quirk, R., Greenbaum, S., Leech, G., Swartvik, J.: A Comprehensive Grammar of the English Language. Longman, Harlow (1985)

19. Sæbø, K.J.: "How"-questions and the manner-method distinction. Synthese **193**, 3169–3194 (2016)

20. Schäfer, M.: Positions and interpretations. German adverbial adjectives at the syntax-semantics interface. Doctoral dissertation, Universität Leipzig (2005)

21. Schäfer, M.: Positions and Interpretations. German adverbial adjectives at the syntax-semantics interface. De Gruyter Mouton, Berlin (2013)

22. Piñón, Ch.: Manner adverbs and manners. Handout, 7. Ereignissemantik-Konferenz, Tübingen, pp. 20–21, December 2007. http://pinon.sdf-eu.org/

23. Pittner, K.: Adverbiale im Deutschen. Stauffenburg Verlag, Tübingen (1999)

24. Van Valin, R.D., Wilkins, D.P.: The case for 'effector': case roles, agents, and agency revisited. In: Shibatani, M., Thompson, S.A. (eds.) Grammatical Constructions, pp. 289–322. Oxford University Press, Oxford (1996)

25. Wyner, A.: Subject-oriented adverbs are thematically dependent. In: Rothstein, S. (ed.) Events and Grammar, pp. 333–348. Kluwer, Dordrecht (1998)

Social Choice and the Problem
of Recommending Essential Readings

Silvan Hungerbühler[1], Haukur Páll Jóhnsson[2], Grzegorz Lisowski[3],
and Max Rapp[4(✉)]

[1] ETH Zurich, Zurich, Switzerland
`silvan.hungerbuehler@bluewin.ch`
[2] SURFsara, Amsterdam, The Netherlands
`haukurpalljonsson@gmail.com`
[3] University of Warwick, Coventry, UK
`grzegorz.lisowski@warwick.ac.uk`
[4] FAU Erlangen-Nürnberg, Erlangen, Germany
`max.rapp@fau.de`

Abstract. We tackle the practical problem of finding a good rule to
recommend a collective set of *news items* to a group of media consumers
with possibly very disparate individual interest in the available items.
For our analysis, we adapt a formal framework from voting theory in
Computational Social Choice to the media setting in order to compare
the performance of five recommendation rules with respect to several
desirable properties of recommendation sets. Through simulations, we
find that polarization of the audience limits how well these rules can
perform in general. On the other hand, greater diversity or universality
can be achieved at only low cost in utility.

Keywords: Computational Social Choice · Recommender systems ·
Multiwinner voting

1   Introduction

How to balance the media's core function of providing news that is relevant to
society at large against the increasing economic necessity of offering an individually tailored product? News media face a dilemma: Either submit to highly
personalized news feeds on online social media networks that drive political
fragmentation, partisanship and contribute to the erosion of society's commonly
accepted factual base; or risk losing disgruntled readers, who feel that the issues
which they consider important are inadequately represented in the mainstream
media, to less reliable Internet news outlets.

We would like to thank Ulle Endriss for guidance and very helpful suggestions. We are
also grateful to anonymous reviewers for detailed and constructive feedback. At the
time at which this work was carried out, all authors were affiliated with the Institute
for Logic, Language and Computation, University of Amsterdam.

© Springer-Verlag GmbH Germany, part of Springer Nature 2019
J. Sikos and E. Pacuit (Eds.): ESSLLI 2018, LNCS 11667, pp. 62–78, 2019.
https://doi.org/10.1007/978-3-662-59620-3_4

Common recommender systems such as matrix factorization algorithms create highly individualized rankings over items based on the users' past behavior [7]. Could those rankings be aggregated in a principled way to generate a common set of *essential readings* for the whole user group?

The present paper takes a step towards addressing this problem by designing and testing a number of such aggregation rules for news articles using the tools of Social Choice Theory. In this approach it is studied how to provide a collective view based on individual opinions fairly. In particular, we might be interested in aggregating individual preference orderings, which is essential for instance in voting mechanisms. Then, intuitive and desired properties (axioms) of social choice rules can be formalised. An extensive overview of the research in this area is provided in [4].

All that is needed for the rule to work in the setting we consider is an ordering of the news articles from first to last according to their importance for each agent. The way in which this preference ordering is elicited from the individual is left open; depending on the concrete application, the data can be thought of as output of a recommender system as suggested above, but could also be explicitly provided by the consumers or gathered by data mining techniques.

Naturally, there are certain properties one would expect such a collection of essential articles to have. The total length of recommended articles for a newspaper's title page, for example, should not exceed its character limit which relates to a problem of making collective choices with a restricted budget [9]. Likewise, there are relations between the essential articles and the rankings by the individuals one would like to see respected by a recommendation rule. For instance, if all consumers detest a certain news item, then it should certainly not be featured in the essential collection instead of another item prioritized by everybody.

Our Contribution. This paper aims at better understanding of collective recommendation rules in media settings by formally studying the interaction between rules and properties of their recommendations. We suggest performance metrics to analyze benefits and drawbacks of various ways to determine a set of essential news items for a group, given each member's individual preferences over said items. The metrics we adopt are based partly on pairwise contest performance of the items and partly on agents' *utilities* over the items, which are derived from their preference orderings. In this way we strive to capture quantitatively the satisfaction particular consumers derive from the recommended selection. A similar, utilitarian approach has been adopted in the social choice literature for instance in [3].

In the current setting, we are in large part interested in how well the recommendation rules perform depending on the structure of the group of consumers that the recommendation is aimed at. It is not difficult to provide a fair recommendation for a group with uniform preferences. However, as the differences between consumers become apparent, making a selection taking into account needs of all consumers becomes challenging. This is why we test the performance of the designed rules for groups diversified in terms of their uniformity.

We proceed by running simulations to estimate the performance of the aggregation rules according to those metrics. We perform two types of analysis. Firstly, we test how rules behave for groups of agents who have similar views against groups with polarized opinions. Secondly, we compare performance of designed rules on the global level.

Related Literature. The proposed approach contributes to the exploration of connections between recommender systems and social choice theory. This direction is well grounded in the literature. One of the lines of research investigating this problem is focused on *collaborative filtering*. There, methods of predicting a preference ordering of a particular user based on preference orderings of a selected group of users are sought. In [11] it was proposed to study desirable properties of collaborative filtering methods originating in social choice theory. Contrary to this approach, in our setting we propose and study the properties of aggregation mechanisms that are employed downstream from such a classical recommender system.

Another related application of social choice methods in recommender systems is in *reputation systems*, in which an ordering of agents (constituting their reputation) is derived from individual orderings that they have over their peers. In [13] it is proposed to investigate the properties of reputation systems analogously to the study of axioms in social choice. Similarly, [1] apply the axiomatic approach to the study of methods of constructing *personalized rankings* for particular agents. It would be interesting to apply the performance metrics and aggregation mechanisms we devise in this paper in such a setting.

Further, the line of work followed in this paper links to recent investigations on social choice rules ensuring the respectable treatment of minorities. For instance, [2] study multi-winner approval voting rules ensuring that sufficiently large groups of voters agreeing that some group of candidates should be elected receive a representative on the chosen committee.

Finally, our research is related to the investigations of the impact of diversity of agents in social choice. For instance, in [6] the effect of diversity of considered preference orderings on the social choice theoretic factors (such as the likelihood of encountering a Condorcet cycle) is studied.

Structure of the Paper. The paper is structured as follows: in Sect. 2 we provide the formal definition of the recommendation problem as we want to study it. In Sect. 3 we propose and formally present desirable properties a collection of recommended articles ought to have. In Sect. 4 we propose five rules for the task of turning individual preferences into a single recommendation. Section 5 contains the methodology, presentation and discussion of our simulation results and suggests directions for future work while Sect. 6 concludes.

2   Formal Framework

This section specifies the formal framework we use. Abstractly, we aim at modeling any setting where a selection of items has to be chosen under a budget constraint for a group of agents with different preferences. More concretely we are interested in the following use cases:

- Traditional media such as newspapers, TV stations, news websites which at any given time exhibit *the same* presentation (modulo individualized advertising and adaptations to screen sizes/devices) of their content to every consumer and face the problem of deciding how to fill the available space or time optimally.
- Non-conventional media such as social networks and news aggregators with highly individualized presentation which aim at reserving a subset of their content space for *essential* items which should be displayed to all of their users regardless of their individual traits.

We assume that in either case a traditional recommender system is already in place providing us with individual rankings over news items. We do not place any restrictions on how this recommender system obtains those rankings - be it through statistical inference from consumers' past behaviour or through direct elicitation of consumer preferences. We do however assume that such a system has tie breaking and other disambiguation mechanisms in place to always be able to give a recommendation. Thus the input to the aggregation mechanisms we propose can be assumed to be a profile of *strict total preference orderings* based on the ranking the recommender system yields after completely disambiguating the preference orderings. This profile is defined as follows: the *news items* are modeled as a set $A = \{a_1, \ldots, a_m\}$, a subset $W \subseteq A$ of which are the *recommended items* for a group of *consumers* $N = \{1, \ldots, n\}$. Each consumer $i \in N$ has preferences over A represented by a strict, total order $\succ_i$[1]. Let $\mathcal{L}(A)$ be the set of such orders over A. Then a vector of preference orders of a set of consumers N over news items A forms a *profile of preferences* $\mathcal{R} \in \mathcal{L}(A)^n$.

In addition, each news item A is assigned a specific *cost* by a function $C : A \to \mathbb{N}$. Depending on the context, the cost of an article could be interpreted as the time it takes to read it, the cognitive resources it takes a consumer to digest it, the space it takes on a newspaper page or website or simply its character length. Notice that the notion of cost can be straightforwardly generalized to sets of articles. Namely, the cost of a recommendation set $W \subseteq A$ is given by:[2]

$$C(W) = \sum_{w \in W} C(w).$$

---

[1] A strict, total order is a transitive, antisymmetric and irreflexive, connected relation.

[2] An anonymous reviewer has pointed out to us that this assumption may be too strict: reading one article may reduce the effort it takes to read another one on the same topic. Thus the cost of reading both articles may well be less than the sum of their individual costs. It would indeed be interesting to generalize the cost function in this way in future investigations.

Finally, we denote by $C^m(A)$ the cost vector $(C(a_1), ..., C(a_m))$.

Let us illustrate these notions with an example.

Example 1. A group of three agents: Alice, Bob and Cindy would like to choose articles that all of them should read. They have three choices: a short item on feminism (*fem*), a long article on the stock exchange (*stock*), and a medium paper on fashion (*fash*). Agents have preferences over the items, which can be represented as follows:

- Alice: *stock* ≻ *fem* ≻ *fash*
- Bob: *fash* ≻ *fem* ≻ *stock*
- Cindy: *fem* ≻ *stock* ≻ *fash*

However, members of the group only have a limited amount of time to read through the papers. So, they are considering the cost of articles. As *fem* is only a short entry, it only takes 10 min to read. So, they assign it a cost of 1. *Fash* is longer, it might take 20 min to read, so they give it a cost of 2. As *stock* is a long and dense paper and reading it can even take 30 min, they assign it the cost of 3.

In addition to the factors defined before, we consider the *utility* that the inclusion of an article in the recommendation set gives to particular consumers. Intuitively, we capture how much particular agents desire to push an object to the recommended set. We follow [9] in deriving pseudo-utilities[3] from consumers' preference orders. For present purposes we used the Borda score, that is, u_i outputs the value $m - 1$ for consumer i's top item, $m - 2$ for the second one and so forth. This choice allows for simplicity of the considered setting. It provides a straightforward conversion of rankings over objects to their utility for users. Other ways of defining utilities are also compatible with our framework. Their exploration and how they affect the outcomes of our simulations would be a possible direction for future research.

The *utility for a consumer i* is given by a function:

$$u_i : A \to \mathbb{N}$$

Similarly to the cost of articles, their utility can also be generalized to sets of items. The *total utility* of a recommendation set W amounts to:

$$u(W) = \sum_{a \in W} u(a)$$

where $u(a) = \sum_{i=1}^{n} u_i(a)$.

Finally, as resources such as attention or the characters on a front page are limited, we assume a *budget* $B \in \mathbb{N}$. In addition, for any $B \in \mathbb{N}$, denote by $\mathcal{W}_B$ the set of all elements W of $\mathcal{P}(A)$ s.t. $C(W) \le B$.

---

[3] They are pseudo-utilities in that they are distinct from the consumers' true utilities that form the input of the upstream recommender system.

Given the notions provided above, we can formulate the definition of a *recommendation rule*. The recommendation rule then is a function from profiles, cost and the budget to recommended items:

$$F : \mathcal{L}(A)^n \times \mathbb{N}^m \times \mathbb{N} \rightarrow \mathcal{P}(A)$$

Let us illustrate the proposed model as a continuation of the example provided before.

Example 2. Recall the previously discussed example. The group has decided that they can only devote 30 min to reading selected articles. So, they can either read a long article, or several shorter papers. Alice wants to read *stock* most, it would give her the utility of 2. *Fem* would get her the utility of 1 and *fash* would not make her happy at all. So, selecting the long article only would give her more benefit (2) than selecting two shorter papers $(1 + 0)$. We can establish to what extent other group members benefit from particular sets of articles analogously. Then, a recommendation rule should provide a group with the "essential" choice of articles within their budget.

In the next section we will turn to the question what "essentiality" might mean in this context. It is worth noting that given a profile of consumers, cost of items and a budget, a recommendation set can be selected in many ways. In order to determine the appropriateness of particular rules we establish a number of *performance metrics*. Each of the performance metrics pertains to a different notion of "essentiality".

3   Performance Metrics

In general, there may be many different views on what exactly renders a news item essential. Should it make the medium's consumers happy? Should it be part of a wide range of content that appeals to a diverse selection of people? Depending on one's stances on these values one may arrive at different conclusions. Thus we do not venture to propose a universally acceptable definition here. Instead, we will present and discuss here a range of desirable properties that reflect different values one might employ to assess the quality of a recommendation set. Practitioners may place different weights on these values and therefore arrive at different trade-offs between them depending on their goals.

The proposed desired properties of functions selecting a number of options from a set, such as those provided by [5], are often binary. A specific function either satisfies them, or it does not. Such properties are standardly referred to as *axioms*. However, following the taken, utilitarian approach, we will be focusing rather on quantitative information about consumers satisfaction. Therefore, in this work we chose to study properties which functions might satisfy to a certain degree. This choice allows for a more robust comparison between performance of different functions. We refer to those properties as *metrics*.

3.1 Utility Maximization

The Utility Maximization metric checks whether a set maximizes provides the highest global utility among the sets whose cost does not exceed the budget. This metric follows the long tradition arguing in favor of utilitarian social welfare functions (e.g. [10]).

Recall that $\mathcal{W}_B$ is the set of all elements of $\mathcal{P}(A)$ s.t. $C(W) \leq B$. A recommendation set W satisfies Utility Maximization iff:

$$W \in \arg\max_{W' \in \mathcal{W}_B}(u(W'))$$

The motivation behind this property is that, arguably, a recommendation set should get consumers the highest possible payoff. For example, highly popular items about viral memes or the latest celebrity scandals would be favored under a rule that maximizes utility. Even if utility is not maximized, it is of interest to assess the gap to optimality. When investigating rules along various other performance dimensions, this allows us to determine the price in terms of utility of improving recommendation sets with respect to those metrics.[4]

3.2 Gini-Coefficient

Utility maximization disregards issues of inequality completely. Thus the utility maximizing recommendation set may be one benefiting just one consumer. In light of this it seems appropriate also to assess recommendation sets in terms of (in)equality. The Gini-coefficient is the most commonly used measure of inequality in a population. There are many equivalent definitions of the Gini-coefficient [14], we define it here in terms of the mean absolute difference between utilities, i.e.:

$$G(W) = \frac{\mathbb{E}[|u_i - u_j|]}{2|W|^{-1}u(W)} = \frac{\frac{1}{|W|^2}\sum_{i=1}^{|W|}\sum_{j=1}^{|W|}|u_i - u_j|}{2|W|^{-1}u(W)} = \frac{\sum_{i=1}^{|W|}\sum_{j=1}^{|W|}|u_i - u_j|}{2|W|u(W)}$$

---

[4] One might wonder at this point whether another utility-related measure, Pareto-efifficiency, would be of use in our setting. In the budgeted context, a winner set W, is Pareto-efficient if there is no combination of items $V = \{v_1, ..., v_n\}$ such that there is $w \in W$ and $i \in N$ with $u_i(V) > u_i(w)$, $u_j(V) \geq u_j(w)$ for all $j \neq i$ and $C(V) \leq C(w)$. In other words, there is no way to replace an item in W such that nobody is worse off, at least one consumer is better off and still fit the budget. To answer this question, one should recall that the utilities employed here are pseudo-utilities: the preference orders on which they are based do not contain information on whether some combination of lower ranked items would actually be preferred over a single higher ranked item. Utilities in our setting merely serve as a way to obtain an aggregate measure of the popularity of a winner set. They do not contain information that would allow one to make between-item or between-consumer comparisons. For these kind of comparisons we need to fall back to the preference orderings. In social choice, the notion corresponding to Pareto-efficiency is called unanimity. Studying unanimity in our multi-winner, budgeted setting would require an adaptation of the usual axiom. We leave this as an interesting avenue for future research.

The Gini-coefficient ranges from 0 (perfect equality, which means that every-body has the same amount of utility) to 1 (perfect inequality, one agent has all the utility). We care about the Gini-coefficient since unequal distributions of utility increase the likelihood that the worst-off consumers lose interest. But desertion of too large a part of the audience would defeat the point of a common recommendation set. Instead the goal might be to keep everyone just happy enough to keep engaged.

For example, construing the recommendation set as the set of characters on a popular TV show, a producer might face the choice, which, if any, of the characters they should kill off. Killing nobody may take away from the show's suspense. Killing too popular a character may outrage the character's fans too the point of losing interest in the show. Instead, hurting everybody a little may just be the best option. Recommendation sets with a low Gini-coefficient are thus arguably preferable to highly unequal ones.

3.3   Condorcet-Based Metrics

The next two metrics are inspired by the influential Condorcet-criterion for assessing voting rules. The intuition behind the Condorcet method is that pair-wise contests should determine winners in an election. An item a is a *Condorcet-winnner* iff it beats every other item b in a majority contest. That is, a is ranked above b on a majority of the preference orders in the profile for every $b \in A$. There are always either one or zero Condorcet-winners. A voting rule satisfies the *Condorcet-criterion* iff it always elects the Condorcet-winner if it exists. *Condorcet-extensions* are rules that satisfy the Condorcet-criterion and elect some other item if the Condorcet-winner does not exist. In a multiwinner setting like ours the question is how to select further winners once the Condorcet winner has been added to the recommendation set. The two metrics presented here are two ways to do this in a way that generalizes the intuition that pairwise contests should determine outcomes.

General Threshold. The first metric generalizes the Condorcet method by not only considering the items that win a majority of the *votes* in each contest but also the ones that win a *qualified minority* of votes. Let $N_{a \succ b}$ denote the set of all consumers who rank a over b. We call $\theta \in [0,1]$ a *general threshold* for a recommendation set W iff

$$a \in W \text{ whenever } \frac{|N_{a \succ b}|}{|N|} \geq \theta \text{ for all } b \in A \setminus \{a\}$$

A recommendation set W is θ-consistent if θ is a general threshold for W. The intuition here is that an item should be in the recommendation set if a qualified minority of the consumers likes the item a lot. The lower the general threshold, the smaller the coalition of consumers needed to predictably push items "on the agenda" as long as they rank those items high enough. If one assumes that an issue's rank corresponds in some sense to how much it affects an individual, the general threshold is a measure of how much importance a recommendation rule assigns to pressing minority issues compared to less salient mainstream topics. For example, given a profile containing a small group of people at high risk of strokes and a large majority of people suffering from mild headaches, a rule that attempts to optimize the general threshold would favor reports on the stroke issue. Note that the general threshold really is Condorcet-consistent: for any profile, if the Condorcet winner c exists, it will be in W and at $\theta = 0.5$, we have that $W = \{c\}$.

Majority Support. Next we consider another generalization of the Condorcet method. Here we care not only about items that win a majority of *pairwise contests* but also about items that win at least a qualified minority of contests.

Note that the number of pairwise majority contests for a given news item is $|N| - 1$. Then $\sigma \in [0, 1]$ is called a majority support (majority support) threshold for a recommendation set W iff

$$a \in W \text{ whenever } \frac{|\{b \in A \setminus \{a\} : \frac{|N_{a \succ b}|}{|N|} > \frac{1}{2}\}|}{|N| - 1} \geq \sigma$$

A recommendation set is σ-consistent if σ is a majority support threshold for W.

In contrast to the general threshold, a low majority support threshold allows consumers to put items on the agenda by establishing a *majority coalition*, even if they do not consider these items as essential. So a rule that minimizes the majority threshold may favor reports on mild headaches over reports on strokes in the previous example.

Again, the Condorcet winner, if it exists, will be in the recommendation set. Furthermore, it should be noted that in the single winner case the majority support rule is known as the Copeland method. In fact, for any σ, any item with a Copeland score (share of majority contests it wins) above σ will also be in W.

Note that utility maximization and Gini-coefficient as well as general threshold and majority support are somewhat complementary: intuitively, a rule which performs well for one of them will have to trade off on the others. "Essentiality" is thus characterized by a point in the space delimited by the poles optimality and equality on one axis and diversity and universality along another. Designing a rule corresponds to choosing such a point and then finding a mechanism that optimizes with respect to this point. This is not an easy task: leaning towards optimality may incur high inequality and desertion of a part of the audience. On the other hand focusing on equality may make for a recommendation set that does not appeal to anyone. Prioritizing universality could lead to neglecting pressing issues that affect only minority groups. Favoring diversity on the other hand risks alienating the average consumer by pushing issues on them which do not affect or interest them.

4   Recommendation Rules

In this section, we study three rules: Lowest General Budget-compatible Threshold (LGBT), Budgeted Utility Maximization, and Budgeted Copeland. These three are designed to perform optimally on the metrics utility maximization, general threshold and majority support, respectively. We will compare them with two more traditional rules: Budgeted Borda and Budgeted Plurality. Budgeted Utility Maximization is equivalent to the well-known Knapsack problem. Budgeted Copeland, Borda and Plurality are budgeted versions of the well-known k-multiwinner voting rules [5]. The LGBT-rule is novel to the best of our knowledge.

4.1   Extending Multiwinner Rules

We chose to adapt three k-multiwinner voting rules proposed by [5] for our setting: A budgeted Plurality rule (Budgeted Plurality) as a baseline, a budgeted Borda rule (Budgeted Borda) as a more sophisticated representative of the positional scoring rules and a budgeted Copeland rule (Budgeted Copeland) to represent the Condorcet extensions. All of these rules assign a score to each item based on the current profile: $S_F : \mathcal{L}(A)^n \times A \to \mathbb{R}$; for brevity we will henceforth refer to all rules that assign a score by *fit-by-score rules* and denote an item's score, given a profile and a rule, as $S_F(a)$, instead of $S_F(\mathcal{R}, a)$.

The *fit-by-score rules* use their respective scores to recommend items in the following way: Start with the complete budget B and put the highest scoring budget-fitting items in the recommendation set. Then do the same for the remaining budget. Continue until the budget is depleted. More formally, recall that the recommendation rule is a function $F : \mathcal{L}(A)^n \times \mathbb{N}^m \times \mathbb{N} \to \mathcal{P}(A)$. Then we define recursively:

$$W_1 = \arg\max_{a \in A} S_F(a) \cap \mathcal{W}_B$$

$$W_k = W_{k-1} \cup (\arg\max_{a \in A \setminus W_{k-1}} S_F(a) \cap \mathcal{W}_{B - C(W_{k-1})})$$

$$F(\mathcal{R}, C^m(A), B) = W_{|A|}$$

4.2   Rules Designed for Optimal Performance

It is easy to see that *fit-by-score* combined with the Copeland-Score performs optimally with respect to majority support. It elects the winning set with the lowest possible majority support threshold σ. Intuitively, this means that it enables majority coalitions to eliminate an item from the agenda even if they dislike that item only a bit. Similarly, we designed rules to perform optimally with respect to utility maximization and general threshold.

Lowest General Budget-Compatible Threshold (LGBT). The first such rule is the *Lowest General Budget-compatible Threshold Rule* (LGBT). It is designed to yield optimal results with respect to general threshold. To achieve this, we start by defining an item's θ-score as follows:

$$\theta(a) = \min_{b \in A \setminus \{a\}} \frac{|N_{a \succ b}|}{|N|}$$

Then LGBT applies the *fit-by-score* method to elect the recommended set. LGBT recommends the set that is optimal for minority preferences in the sense that it chooses the recommendation set with the lowest possible general threshold, thus enabling comparatively small coalitions to push the issues they consider important onto the agenda. For this it is sufficient but not necessary that a plurality of $\theta|N|$ voters rank the desired item first.

Example 3. Recall the example discussed before. There, the article *fem* is preferred to *stock* by Bob and Cindy. It is also preferred to *fash* by everybody. So, it's θ-score amounts to $\frac{2}{3}$. Similarly, we can compute that the θ-score of *stock* is $\frac{1}{3}$ and *fem* receives $\frac{2}{3}$. Hence, the LGBT rule will select *fem* first (assuming the tie between *fem* and *fash* is broken in favor of the item with lower cost). Next *fash* is added. As then the budget is exhausted, the recommendation set will contain *fem* and *fash* and have a general threshold of $\frac{2}{3}$.

Budgeted Utility Maximization. The second novel recommendation rule is the *Budgeted Utility Maximization*. Budgeted Utility Maximization is designed to perform optimally with respect to utility maximization. It selects the set of articles which has the greatest overall utility for the consumers while fitting the budget.

$$F(\mathcal{R}, C^m(A), B) = \arg\max_{W \in \mathcal{W}_\mathcal{B}} u(W)$$

Example 4. Recall the example discussed before. There, the article *fem* would give 1 utility point to Alice and to Bob, as well as 2 points for Cindy. Then, the paper *stock* would generate 2 points for Alice, 1 point for Cindy and none for Bob. Finally, *fash* would only generate 2 points for Bob. Due to the budget restriction, we can either choose only *stock*, *fash* and *fem* or only one of them. Selecting *fash* and *fem* would clearly maximize the global utility generated by the allowed sets. Thus, it will be selected by the Budgeted Utility Maximization rule.

Although computing the Budgeted Utility Maximization recommendation is an NP-hard problem because it corresponds to the *0–1 Knapsack Problem*, there exist algorithms to solve it in pseudo-polynomial time [8]. We used such an algorithm for our simulations.

5   Simulations

We have already established that *fit-by-score* with the Copeland score is optimal with respect to the majority support metric, LGBT with respect to the general threshold metric and Budgeted utility maximization rule with respect to utility maximization[5]. However, this does not tell us yet how much of a price each rule has to pay to achieve optimality on its respective performance metric. Maybe the cost in equality, diversity and universality is typically low when one maximizes utility? Or it might be the case that LGBT achieves only a marginal improvement in terms of diversity over the other rules at a high cost in utility? In addition, the distribution of individual preference orders in a profile possibly affects the performance of each aggregation rule. Some rule could function well for very homogeneous preferences, while performing poorly when preferences are very fragmented or even polarly oppose each other. There is no clear way to answer these questions other than empirically. Since only a very limited number of naturally occurring preference profile datasets are available, we chose to proceed experimentally through computer simulations.

5.1   Method

We generated profiles ourselves in order to capture some plausible types of distributions of preferences amongst a group of individuals. Thereafter we automatically checked the performance of the considered rules with respect to the four desirable properties: Utility Maximization, Gini-Coefficient, General Threshold and Majority Support. The performance was compared with respect to specific profiles.

In order to generate preference profiles as they might naturally occur in different types of consumer populations, we specified seven different *base profiles*. Each of them represented a possible distribution of individual preferences over 10 and 20 items respectively. Every profile contained preference orders of 5000 individual consumers.

The seven base preference orders fall in four categories: *random* profiles—each individual's preference order is sampled independently from the other 4999 preferences in the profile; *fraction* profiles—there are five different clusters of individual preferences which are more similar to each their clustered peers than to the other clusters (there are, in turn, two types of cluster-size distributions: one where the biggest cluster comprises 50% of individual orders and five smaller ones 10% each, and another one where the division is 80% and four times 5%); *polarized* profiles—where two consumer populations have polarly opposed preference orders; and *similar* profiles—where two consumer populations have different preferences orders, but not in a polarized way.

---

[5] One might ask why we did not consider a rule that minimizes the Gini-coefficient. The reason is that usually even proponents of egalitarianism don't aspire to minimize inequality (in all domains) but rather put side constraints on utility maximization, as in the famous *Maximin*-principle [12]. The Maximin-principle would correspond to the rule $F(\mathcal{R}, C^m(A), B) = \arg\max_{W \in \mathcal{W}_\mathcal{B}} \min_{i \in N} u_i(W)$ in our setting.

A specific profile, then, is a noisy copy of a base profile. After specifying the base profiles we applied noise to model variety amongst individual consumers. The noise was introduced by employing a probabilistic model to swap items in the preference orders, where farther swaps occurred with smaller probability than closer swaps. Namely, a rank r in the preference order and a swap direction (upwards or downwards) are sampled random-uniformly. Then for each rank r' lower (or higher respectively) than the sampled rank, a swap of the corresponding items is attempted. The success probability of the swap is:

$$\frac{1}{(|r - r'| + 1)^2}$$

This way of introducing noise allows to control the swapping distance between profiles within a cluster by varying the number of ranks to be sampled. It captures the intuition that within a cluster, profiles should be similar to each other. In this fashion we generated 100 profiles each for the cases of 10 and 20 news items for all seven profile-types.

In terms of the results presented below, the most important base profiles are the following two with two clusters of underlying preferences each. They are meant to account for the differences in the performance of rules in situations when consumers' opinions are homogeneous, and when they are diversified.

For the first, the two clusters are simply noisy copies of a third underlying preference order. We randomly generated one preference order, applied high noise to it twice to get two different preference orders and then generated 2500 moderately noisy copies of both. This resulted in profiles with two *clusters* of fairly similar consumer groups whose orderings agree mostly, but not completely, amongst themselves. We call them *similar-cluster profiles*. For the other base profile with two underlying preference orders, we randomly generated a preference order and created 2500 noisy copies of it. We then reversed the initial ordering and generated 2500 noisy copies of it as well. We call the result *polarized-cluster profiles*.

5.2   Results

We present here our most salient results which pertain to (a) how different profiles affect performance on average with respect to our desired properties *for all of our rules*. This test allowed for checking what is the difference between the performance of rules in uniform and polarized societies. Further, results of type (b) show how the rules perform *against each other* on the generated profiles. This allowed us to test for differences across distinct profiles for each rule. To test the significance of the obtained results, we employed ANOVA in combination with Tukey post hoc analysis.

Results of Type (a). Firstly, concerning utility maximization, we found that for all profiles with 20 candidates and two clusters of consumers, the rules performed on average 5 to 10% points worse on polarized-clusters populations than on similar-clusters populations.

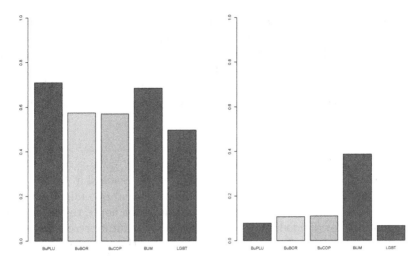

Fig. 1. From left to right the bars stand for Budgeted Plurality, Budgeted Borda, Budgeted Copeland, Budgeted Utility Maximization and LGBT. The left plot shows the average general threshold for polarized profiles. The right plot shows average general threshold for similar profiles.

Secondly, with respect to general threshold and profiles with two clusters of consumers, all rules perform between 30 and 55% points worse on the polarized-cluster populations than on the similar-cluster populations. This can be seen in Fig. 1. For LGBT, these differences were significant on a 95% confidence level for both the 10 and 20 items case. For Budgeted Borda, Budgeted Copeland and Budgeted Plurality they were only significant for the profiles with 10 items.

Results of Type (b). We found that both Budgeted Borda and Budgeted Plurality perform significantly worse with respect to utility maximization than Budgeted Utility Maximization on the 95% confidence level. While Budgeted Copeland and LGBT performed worse on utility maximization on average, the difference to Budgeted Utility Maximization was not significant.

With respect to general threshold, we found that for both for the populations with 10 and 20 items, the LGBT rule performs significantly better than the other rules on the 99%-level. The difference is most pronounced for the Budgeted Utility Maximization rule. For 10 items Budgeted Utility Maximization scores on average 19.5% points higher (worse) than LGBT. In the case of 20 items, Budgeted Utility Maximization scores on average 21.8% points higher (worse) than LGBT.

For the Gini-coefficient we found that for 20 and 10 items, respectively, Budgeted Utility Maximization performs 3.8 and 10.6% points worse than LGBT, with significance on the 99% level. There were no significant differences between LGBT and the other score-and-fit rules.

Another result pertains to majority support. As expected, Budgeted Copeland performed best with respect to this metric and, significantly so, compared to all other rules on the 99% interval. Less obviously, we also found that Budgeted Borda outperforms the remaining rules with a significance on the 99% level. Thus, both Budgeted Utility Maximization and LGBT only did approximately as well as the baseline Budgeted Plurality rule with respect to majority support.

5.3   Discussion

The (a)-type results indicate that polarization of the readership leads to both lower utility and higher general threshold indicating what one might call a *cost of polarization*. The cost of polarization is bad news for hopes to find common recommendations for divided audiences. An avenue for future research could be the search for a recommendation rule that minimizes the cost of polarization.

Concerning type (b) results, we conclude that no single rule performs optimally with respect to all desirable properties and types of profiles. As expected we have a "three-way-tie" between Budgeted Utility Maximization, LGBT and Budgeted Copeland when it comes to utility maximization, general threshold and majority support. However, looking at the other properties, a tendency can be established: While the Budgeted Utility Maximization rule performs optimally on utility maximization, the difference to Budgeted Copeland`and LGBT was not significant which indicates that Budgeted Copeland and LGBT come close to the optimum. A further point to make is that while LGBT and Budgeted Copeland are tied with respect to Gini-coefficient, they clearly beat Budgeted Utility Maximization in this respect.

To sum up: although Budgeted Utility Maximization maximizes utility by construction, this only leads to little added utility in comparison with Budgeted Copeland and LGBT. Moreover, this additional utility comes at the cost of markedly unequal utility distribution amongst the consumers, higher lowest general threshold and higher lowest majority support thresholds. All of this arguably indicates that Budgeted Utility Maximization is overall a worse rule than Budgeted Copeland and LGBT: higher diversity or greater universality can be achieved at a fairly low price in overall utility. This indicates that greater individualization is not the only way forward when attempting to capture and maintain an audience. Thus maybe the media are not doomed to ever greater splintering and catering to special interests if essential readings are chosen prudently.

In the present paper, we have suggested and tested several rules to carry out this choice. However, in doing so we made some strong assumptions, most notably the equi-distance between neighboring items' utilities. In addition, our results are based purely on simulations. Hence it would be a natural next step to investigate whether our simulation results carry over to an experimental setting with utility data submitted by real media consumers, such as the users of an online news-aggregator. In this setting, one could then also track how a common set of essential readings affects users' preference orders over time. This would also allow to research strategic behavior in the considered framework. Then, we might identify recommendation rules which are the hardest to manipulate.

6   Conclusion

In this paper, we presented a formal Social Choice framework to recommend a common winner set to a group of consumers given a budget constraint. Interpreting the items as news articles, the voters as readers and the budget as e.g. readers' attention span or space on a front-page this task can be understood as finding a principled way to balance individualization and newsworthiness by selecting a set of *essential readings* to be recommended to all readers. We devised novel performance metrics suitable to this setting which provide different ways to evaluate the recommended set. With these criteria in mind, we introduced five rules, one of them novel, designed to perform especially well. For all of the rules, we ran simulations in order to assess the rules against our performance measures. For the simulations we defined multiple population-types and for each type, we generated multiple profiles. We then applied our voting rules to the profiles and performed statistical analysis on the results.

Our conclusion is that using Social Choice theory offers an interesting avenue to improve upon existing recommender systems. Depending on one's value judgments, Budgeted Copeland, LGBT and Budgeted Utility Maximization perform well at generating principled common recommendations from individual preference orders. Notably however, the Budgeted Utility Maximization rule only achieves marginal improvement in utility at the expense of diversity, universality and equality. Thus we conclude that LGBT and Budgeted Copeland are the best rules in this setting known so far.

On the other hand we found that polarization of the audience limits what can be achieved by any of these rules. The cost of polarization thus presents an open challenge for designing recommendation rules in the presented setting.

References

1. Altman, A., Tennenholtz, M.: An axiomatic approach to personalized ranking systems. J. ACM (JACM) **57**(4), 26 (2010)
2. Aziz, H., Brill, M., Conitzer, V., Elkind, E., Freeman, R., Walsh, T.: Justified representation in approval-based committee voting. Soc. Choice Welf. **48**(2), 461–485 (2017)
3. Boutilier, C., Caragiannis, I., Haber, S., Lu, T., Procaccia, A.D., Sheffet, O.: Optimal social choice functions: a utilitarian view. Artif. Intell. **227**, 190–213 (2015)
4. Brandt, F., Conitzer, V., Endriss, U., Lang, J., Procaccia, A.D.: Handbook of Computational Social Choice. Cambridge University Press, New York (2016)
5. Elkind, E., Faliszewski, P., Skowron, P., Slinko, A.: Properties of multiwinner voting rules. Soc. Choice Welf. **48**(3), 599–632 (2017)
6. Hashemi, V., Endriss, U.: Measuring diversity of preferences in a group. In: ECAI, pp. 423–428 (2014)

7. Jannach, D., Resnick, P., Tuzhilin, A., Zanker, M.: Recommender systems – beyond matrix completion. Commun. ACM **59**(11), 94–102 (2016)
8. Kellerer, H., Pferschy, U., Pisinger, D.: Knapsack Problems. Springer, Heidelberg (2004). https://doi.org/10.1007/978-3-540-24777-7
9. Lu, T., Boutilier, C.: Budgeted social choice: from consensus to personalized decision making. IJCAI **11**, 280–286 (2011)
10. Mill, J.: Utilitarianism. Longmans, Green, Reader & Dyer, London (1874)
11. Pennock, D.M., Horvitz, E., Giles, C.L., et al.: Social choice theory and recommender systems: analysis of the axiomatic foundations of collaborative filtering. In: AAAI/IAAI, pp. 729–734 (2000)
12. Rawls, J.: A Theory of Justice. Harvard University Press, Cambridge (2009)
13. Tennenholtz, M.: Reputation systems: an axiomatic approach. In: Proceedings of the 20th conference on Uncertainty in artificial intelligence, pp. 544–551. AUAI Press (2004)
14. Yitzhaki, S., Schechtman, E.: The Gini Methodology: A Primer on a Statistical Methodology. Springer Series in Statistics. Springer, New York (2013). https://doi.org/10.1007/978-1-4614-4720-7

Towards a 2-Multiple Context-Free Grammar for the 3-Dimensional Dyck Language

Konstantinos Kogkalidis and Orestis Melkonian[✉]

Utrecht University, Utrecht, The Netherlands
{k.kogkalidis,o.melkonian}@students.uu.nl

Abstract. We discuss the open problem of parsing the Dyck language of 3 symbols, D^3, using a 2-Multiple Context-Free Grammar. We attempt to tackle this problem by implementing a number of novel meta-grammatical techniques and present the associated software packages we developed.

Keywords: Dyck language ·
Multiple context-free grammars (MCFG) · Young Tableaux ·
Spider webs · Meta-grammars

1 Introduction

Multidimensional Dyck languages [8] generalize the well-known pattern of well-bracketed pairs of parentheses to k-symbol alphabets. Our goal in this paper is to study the 3-dimensional Dyck language D^3, and the question of whether this is a 2-dimensional multiple context-free language, 2-MCFL.

For brevity's sake, this section only serves as a brief introductory guide towards relevant papers, where the interested reader will find definitions, properties and various correspondences of the problem.

1.1 Preliminaries

We use D^3 to refer to the Dyck language over the lexicographically ordered alphabet $a < b < c$, which generalizes well-bracketed parentheses over three symbols. Denoting with $\#x(w)$ the number of occurrences of symbol x within word w, a word belongs in D^3 if and only if it satisfies the following conditions:

(D1) $\#a(w) = \#b(w) = \#c(w)$
(D2) $\#a(v) \geq \#b(v) \geq \#c(v)$, $\forall\, v \in \mathit{PrefixOf}(w)$

Eliding the second condition (D2), we get the MIX language, which models free word order over the same alphabet. MIX has already been proven expressible by a 2-MCFG [14]; the class of multiple context-free grammars that operate on pairs of strings [15].

J. Sikos and E. Pacuit (Eds.): ESSLLI 2018, LNCS 11667, pp. 79–92, 2019.
https://doi.org/10.1007/978-3-662-59620-3_5

1.2   Motivation

Static Analysis. Interestingly, the 2-symbol Dyck language (D^2) is used in the *static analysis* of programming languages, where a large number of analyses are formulated as *language-reachability* problems [12].

For instance, when considering interprocedural calls as part of the source language, high precision can only be achieved by examining only control-flow paths that respect the fact that a procedure call always returns to the site of its current caller [13]. By associating the program point *before* a procedure call f_k with $(_k$, and the one *after* the call with $)_k$, the validity problem is reduced to recognizing D^2 words.

Alas, the 2-dimensional case cannot accommodate richer control-flow structures, such as exception handling via `try/catch` and Python generators via the `yield` keyword. To achieve this, one must lift the Dyck-reachability problem to a higher dimension which, given the computational cost that context-sensitive parsing induces, is currently prohibited. If D^3 is indeed a 2-MCFL, parsing it would become computationally attainable for these purposes and eventually allow scalable analysis for non-standard control-flow mechanisms by exploiting the specific structure of analysed programs, as has been recently done in the 2-dimensional case [1].

Last but not least, future research directions will open up in a multitude of analyses that are currently restrained to two dimensions, such as *program slicing*, *flow-insensitive points-to analysis* and *shape approximation* [12].

Linguistics. For the characterization of natural language grammars, the extreme degree of 'scrambling' permitted by the MIX language may be considered overly expressive [4]. On the other hand, the prefix condition of D^3 allows for partial word movement, while still respecting certain linear order constraints, as observed in natural languages.

Supported by the fact that the language of well-bracketed parentheses, D^2, is a simple CFL (i.e. 1-MCFL) and given that MIX itself is a 2-MCFL, it is reasonable to examine whether D^3 can also be modelled by a 2-MCFG. Such an endeavour proved quite challenging, necessitating careful study of correspondences with other mathematical constructs.

1.3   Correspondences

Young Tableaux. A standard Young Tableau is defined as an assortment of n boxes into a ragged (or jagged, i.e. non-rectangular) matrix containing the integers 1 through n and arranged in such a way that the entries are strictly increasing over the rows (from left to right) and columns (from top to bottom). Reading off the entries of the boxes, one may obtain the *Yamanouchi* word by placing (in order) each character's index to the row corresponding to its lexicographical ordering.

In the case of D^3, the Tableau associated with these words is in fact *rectangular* of size $n \times 3$, and the length of the corresponding word (called a *balanced or dominant Yamanouchi word* in this context) is $3n$, where n is the number of occurrences of each unique symbol [8]. Practically, the rectangular shape ensures constraint (D1), while the ascending order of elements over rows and columns ensures constraint (D2). In that sense, a rectangular standard Young tableau of size $n \times 3$ is, as a construct, an alternative way of uniquely representing the different words of D^3. We present an example tableau in Fig. 1.

a:	1	3	4	8	9	10
b:	2	5	7	11	13	15
c:	6	12	14	16	17	18

Fig. 1. Young Tableau for *"abaabcbaaabcbcbccc"*

Promotions and Orbits. There is an interesting transformation on Young Tableaux, namely the *Jeu-de-taquin* algorithm. When operating on a rectangular tableau $T(n, 3)$, Jeu-de-taquin consists of the following steps:

(1) Reduce all elements of T by 1 and replace the first item of the first row with an empty box $\Box(x, y) := (1, 1)$.
(2) While the empty box is not at the bottom right corner of T, $\Box(x, y) \neq (n, 3)$, do:
  – Pick the minimum of the elements directly to the right and below the empty box, and swap the empty box with it. $T(x, y) := min(T_{(x+1,y)}, T_{(x,y+1)})$, $\Box(x', y') := (x+1, y)$ (in the case of a right-swap) or $\Box(x', y') := (x, y + 1)$ (in the case of a down-swap).
(3) Replace the empty box with $3n$.

The tableau obtained through Jeu-de-taquin on T is called its promotion $p(T)$. We denote by $p^k(T)$, k successive applications of Jeu-de-taquin. It has been proven that $p^{3n}(T) = T$ [10]. In other words, the promotion defines an equivalence class, which we name an *orbit*, which cycles back to itself. Orbits dissect the space of D^3 into disjoint sets, i.e. every word w belongs to a particular orbit, obtained by promotions of T_w.

A_2 *Combinatorial Spider Webs.* The A_2 *irreducible combinatorial spider web* is a directed planar graph embedded in a disk that satisfies certain conditions; we refer the reader to [5] for a formal definition. Spider webs can be obtained through the application of a set of rules, known as the *Growth Algorithm* [10]. These operate on pairs of neighbouring nodes, collapsing them into a singular intermediate node, transforming them into a new pair or eliminating them altogether. Growth rules will be examined from a grammatical perspective in Sect. 2.2. Upon reaching a fixpoint, the growth process produces a well-formed Spider Web, which, in the context of D^3, can be interpreted as a visual representation of parsing a word [8,10].

A bijection also links Young Tableaux with Spider Webs. More specifically, the act of promotion is isomorphic to a combinatorial action on spider webs, namely *web rotation* [10].

Constrained Walk. A Dyck word can also be visualized as a constrained *walk* within the first quadrant of $\mathbb{Z}^2$. We can assign each alphabet symbol x a vector value $\boldsymbol{v_x} \in \mathbb{Z}^2$ such that all pairs of $(\boldsymbol{v_x}, \boldsymbol{v_y})$ are linearly independent and:

$$\boldsymbol{v_a} + \boldsymbol{v_b} + \boldsymbol{v_c} = \mathbf{0} \tag{1}$$
$$\kappa\boldsymbol{v_a} + \lambda\boldsymbol{v_b} + \mu\boldsymbol{v_c} \geq \mathbf{0}, (\forall \; \kappa \geq \lambda \geq \mu) \tag{2}$$

We can then picture Dyck words as routes starting from $(0,0)$. (1) means that each route must also end at $(0,0)$ ($\cong$ (D1)), while (2) means that the x and y axes may never be crossed ($\cong$ (D2)). An example walk is depicted in Fig. 2.

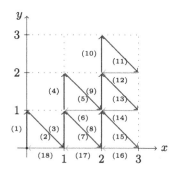

Fig. 2. The constrained walk of *"abaabcbaaabcbcbccc"* with vector value assignments $\boldsymbol{v_a} = (1,0)$, $\boldsymbol{v_b} = (-1,1)$, $\boldsymbol{v_c} = (0,-1)$

2 Abstract Grammar Specification

We now present a number of novel techniques that we developed as an attempt to solve the problem at hand, incrementally moving towards more complex and abstract grammars. For the purpose of experimentation we have implemented these techniques, based on a software library for parsing MCFGs [7]. The resulting Python code is open-source and available online[1].

2.1 Triple Insertion

To set things off, we start with the grammar of *triple insertion* in Fig. 3. This grammar operates on non-terminals $W(x,y)$, producing $W(x',y')$ with an additional triplet a, b, c that respects the partial orders $x < y$ and $a < b < c$. The end-word is produced through the concatenation of (x,y).

---

[1] https://github.com/omelkonian/dyck.

$$S(xy) \leftarrow W(x,y). \tag{1}$$
$$W(\epsilon, xy\mathbf{abc}) \leftarrow W(x,y). \tag{2}$$
$$W(\epsilon, x\mathbf{a}y\mathbf{bc}) \leftarrow W(x,y). \tag{3}$$
$$W(\epsilon, x\mathbf{ab}y\mathbf{c}) \leftarrow W(x,y). \tag{4}$$

$$\cdots$$

$$W(\mathbf{abc}xy, \epsilon) \leftarrow W(x,y). \tag{61}$$
$$W(\epsilon, \mathbf{abc}). \tag{62}$$
$$W(\mathbf{a}, \mathbf{bc}). \tag{63}$$
$$W(\mathbf{ab}, \mathbf{c}). \tag{64}$$
$$W(\mathbf{abc}, \epsilon). \tag{65}$$

Fig. 3. Grammar of triple insertions

Despite being conceptually simple, this grammar consists of a large number of rules. Its expressivity is also limited; the prominent weak point is its inability to manage the effect of *straddling*, namely the generation of words whose substituents display complex interleaving patterns. Refer to Fig. 10 for an example.

2.2 Meta-grammars

To address the issue of rule size, we employ the notion of *meta-grammars*, loosely inspired by Van Wijngaarden's work [16], which allows a more abstract view of the grammar as a whole. Meta-grammars have found wide use in describing linguistic phenomena involving discontinuity and word permutation, giving rise to abstraction techniques that alleviate the need to make all orderings explicit, such as partially ordered multi-set CFGs [9], phrase structure grammars [2,3] and the domain union operation [11].

Specifically, we define $\mathcal{O}$ as the *meta-rule* which, given a rule format, a set of partial orders (over the tuple indices of its premises and/or newly added terminal symbols), and the MCFG dimensionality, automatically generates all the order-respecting permutations. An example of how we can abstract away from explicitly enumerating the entirety of our initial rules is showcased in Fig. 4.

$$S(xy) \leftarrow W(x,y).$$
$$\mathcal{O}_2[\![W \leftarrow \epsilon \mid \{a < b < c\}]\!].$$
$$\mathcal{O}_2[\![W \leftarrow W \mid \{x < y, \ a < b < c\}]\!].$$

Fig. 4. $\mathcal{G}_0$: Meta-grammar of triple insertions

This approach enhances the potential expressivity of our grammars as well. For instance, we can now extend the previous grammar with a single meta-rule that allows two non-terminals $W(x, y)$, $W(z, w)$ to interleave with one another, producing rearranged tuple concatenations and allowing some degree of straddling to be generated:

$$\mathcal{G}_1 : \mathcal{G}_0 + \mathcal{O}_2[\![W \leftarrow W, W \mid \{x < y, \; z < w\}]\!].$$

The addition of this rule gets us closer to completeness, but we are still not quite there. We have thus far only used a single non-terminal, not utilizing the expressivity that an MCFG allows. To that end, we propose new non-terminals to represent incomplete word *states*; that is, words that either have an extra symbol or miss one. The former are *positive* states, whereas the latter are *negative*. The inclusion of these extra states would allow for more intricate interactions, in line with the CFG that describes the 2-letter MIX language [6].

Interestingly, there is a direct correspondence between these non-terminals and the nodes of Petersen's growth algorithm [10]. Figure 5 depicts the growth rules in the exact same web form as proposed by Petersen, modulo node branding. Briefly, these describe a way of constructing minimal parses of D^3 words, as follows:

1. each character in the original string gets translated into a positive state
2. each growth rule combines adjacent elements on the top, either producing one/two new outputs on the bottom or cancelling them out
3. the growth process terminates (i.e. the word is parsed) when there remain no dangling strands

A subset of these web-reduction rules are, in fact, precisely modelled by the meta-grammar $\mathcal{G}_2$ presented in Fig. 6. In Sect. 4, we briefly explain our inability to model the whole set of rules with a 2-MCFG, which would render our grammar complete.

$\mathcal{G}_2$ consists of base cases for positive states, possible state interactions, closures of pairs of inverse polarity and a universally quantified meta-rule that allows the combination of any incomplete state with a well-formed one (i.e. non-terminal W).

A further extension can be achieved through universally quantifying the notion of triple insertion, which is unique in the sense that it can insert three different terminals, each at a different position:

$$\mathcal{G}_3 : \mathcal{G}_2 + \forall \, K \in \{A^{+/-}, \; B^{+/-}, \; C^{+/-}\} : \mathcal{O}_2[\![K \leftarrow K \mid \{x < y, \; a < b < c\}]\!].$$

2.3   Rule Inference

The improved performance of the above approaches again proved insufficient to completely parse D^3. Our meta-rules are over-constrained by imposing a total order on the tuple elements, due to their inability to keep track of where the

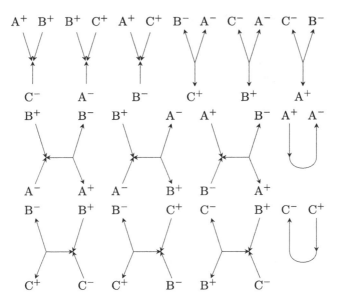

Fig. 5. Growth rules

extra character(s) is. To overcome this, we split each state into multiple position-aware, *refined* states. Doing so revealed a vast amount of new interactions, as evidenced by the below alteration to the original A^+, B^+ interaction (where y can now occur after z or w):

$$\mathcal{O}_2[\![C^- \leftarrow A^+_{left}, B^+ \mid \{x < y, x < z < w\}]\!].$$

In order to accommodate the interactions between this increased number of states, we need to keep track of both internal and external order constraints. At this point, the abstraction offered by our meta-grammar approach does not cover our needs anymore. The same difficulty that we had encountered before is prominent once more, except now at an even higher level.

As a solution to the aforementioned limitation, we propose a system that can automatically create a full-blown m-MCFG given only the states it consists of. To accomplish this, we assign each state a unique *descriptor* that specifies the content of its tuple's elements. Aligning these descriptors with the tuple, we can then infer the descriptor of the resulting tuple of every possible state interaction. For the subset of those interactions whose resulting descriptor is matched with a state, we can now automatically infer the rule.

Formally, the system is initialized with a map $\mathcal{D}$, such as the one illustrated in Fig. 7. Its domain, $dom(\mathcal{D})$, is a set of *state identifiers* and its codomain, $codom(\mathcal{D})$, is the set of their corresponding *state descriptors*.

$$S(xy) \leftarrow W(x, y).$$
$$\mathcal{O}_2[\![W \leftarrow \epsilon \mid \{a < b < c\}]\!].$$
$$\mathcal{O}_2[\![A^+ \leftarrow \epsilon \mid \{a\}]\!].$$
$$\mathcal{O}_2[\![B^+ \leftarrow \epsilon \mid \{b\}]\!].$$
$$\mathcal{O}_2[\![C^+ \leftarrow \epsilon \mid \{c\}]\!].$$
$$\mathcal{O}_2[\![C^- \leftarrow A^+, B^+ \mid \{x < y < z < w\}]\!].$$
$$\mathcal{O}_2[\![B^- \leftarrow A^+, C^+ \mid \{x < y < z < w\}]\!].$$
$$\mathcal{O}_2[\![A^- \leftarrow B^+, C^+ \mid \{x < y < z < w\}]\!].$$
$$\mathcal{O}_2[\![A^+ \leftarrow C^-, B^- \mid \{x < y < z < w\}]\!].$$
$$\mathcal{O}_2[\![B^+ \leftarrow C^-, A^- \mid \{x < y < z < w\}]\!].$$
$$\mathcal{O}_2[\![C^+ \leftarrow B^-, A^- \mid \{x < y < z < w\}]\!].$$
$$\mathcal{O}_2[\![W \leftarrow A^+, A^- \mid \{x < y < z < w\}]\!].$$
$$\mathcal{O}_2[\![W \leftarrow C^-, C^+ \mid \{x < y < z < w\}]\!].$$
$$\forall\, K \in \{A^{+/-},\ B^{+/-},\ C^{+/-}\}:$$
$$\mathcal{O}_2[\![K \leftarrow K, W \mid \{x < y,\ z < w\}]\!].$$

Fig. 6. $\mathcal{G}_2$: Meta-grammar of incomplete states

$W \mapsto (\epsilon, \epsilon)$	$A_r^- \mapsto (\epsilon, bc)$
$A_l^+ \mapsto (a, \epsilon)$	$A_{lr}^- \mapsto (b, c)$
$A_r^+ \mapsto (\epsilon, a)$	$B_l^- \mapsto (ac, \epsilon)$
$B_l^+ \mapsto (b, \epsilon)$	$B_r^- \mapsto (\epsilon, ac)$
$B_r^+ \mapsto (\epsilon, b)$	$B_{lr}^- \mapsto (a, c)$
$C_l^+ \mapsto (c, \epsilon)$	$C_l^- \mapsto (ab, \epsilon)$
$C_r^+ \mapsto (\epsilon, c)$	$C_r^- \mapsto (\epsilon, ab)$
$A_l^- \mapsto (bc, \epsilon)$	$C_{lr}^- \mapsto (a, b)$

Fig. 7. Map $\mathcal{D}$ for refined states

Meta-grammars accelerated the process of creating grammars, by letting us simply describe rules instead of explicitly defining them. ARIS builds upon this notion to raise the level of abstraction even further[2]; one needs only specify a grammar's states and its descriptors, thus eliminating the need to define rules or even meta-rules

---

[2] An avid reader will notice the use of the $\mathcal{O}$ meta-rule in the ARIS algorithm.

Algorithm 1. ARIS: Automatic Rule Inference System

---

procedure ARIS($\mathcal{D}$)
    for $X \mapsto (d_1, \ldots, d_n) \in \mathcal{D}$ **do**
        yield $X(d_1, \ldots, d_n)$.
    for $X, Y \in dom(\mathcal{D})^2$ **do**
        $(X_{ord}, \ Y_{ord}) \leftarrow (x < y < \ldots, \ z < w < \ldots)$
        for $(d_1, ..., d_n) \in \mathcal{O}_2[\![\_ \leftarrow X, Y \mid \{X_{ord}, Y_{ord}\}]\!]$ **do**
            for $S' \in$ ELIMINATE$((d_1, \ldots, d_n), \mathcal{D})$ **do**
                yield $S'(d_1, \ldots, d_n) \leftarrow X, Y$.

procedure ELIMINATE$((d_1, \ldots, d_n), \mathcal{D})$
    for $matches \in$ ALL_ABC_TRIPLETS$(d_1, \ldots, d_n)$ **do**
        for $i \in 0 \ldots n/3$ **do**
            for $S' \in$ REMOVE_ABC_TRIPLETS$(matches, i)$ **do**
                if $S' \in codom(\mathcal{D})$ **then**
                    yield S'

We use ARIS instantiated with map $\mathcal{D}$ of Fig. 7 to generate $\mathcal{G}_4$ as our last attempt to parse D^3. Even though map $\mathcal{D}$ consists of a mere amount of 16 mappings, it produces a lavish parsing system of 1456 concrete rules; disappointingly, these again do not yield a complete solution to our problem.

3 Tools and Results

3.1 Grammar Utilities

We have implemented the modelling techniques described in Sect. 2 and distributed a Python package, called **dyck**, which provides the programmer with a *domain-specific language* close to this paper's mathematical notation. To facilitate experimentation, our package includes features such as grammar selection, time measurements, word generation and soundness/completeness checking. The following example demonstrates the definition of $\mathcal{G}_3$:

```
from dyck import *
G3 = Grammar(initial='W',
  # Base Cases
  O('W', {(a, b, c)}),
  O('A-', {(b, c)}), O('B-', {(a, c)}), O('C-', {(a, b)}),
  O('A+', {(a,)}),   O('B+', {(b,)}),   O('C+', {(c,)}),
  # Combinations
  O('C- <- A+, B+', {(x, y, z, w)}),
  O('B- <- A+, C+', {(x, y, z, w)}),
  O('A- <- B+, C+', {(x, y, z, w)}),
  O('C+ <- B-, A-', {(x, y, z, w)}),
  O('B+ <- C-, A-', {(x, y, z, w)}),
  O('A+ <- C-, B-', {(x, y, z, w)}),
  forall(all_states,
        lambda K: O('K <- K, W', {(x, y), (z, w)})),
  # Closures
  O('W <- A+, A-', {(x, y, z, w)}),
  O('W <- C-, C+', {(x, y, z, w)}),
  # Universal Triple Insertion
  forall(all_states,
        lambda K: O('K <- K', {(x, y), (a, b, c)})))
```

3.2   Visualization

As counter-examples began to grow in size and number, we realised the necessity of a visualization tool to assist us in identifying properties they may exhibit. To that end, we distribute another Python package, called **dyckviz**, which allows the simultaneous visualization of tableau-promotion and web-rotation (grouped in their corresponding equivalence classes). An example of a web as rendered by our tool is given in Fig. 8.

Young tableaux in an orbit are colour-grouped by their column indices, which sheds some light on how the *jeu-de-taquin* actually influences the structure of the corresponding Dyck words. Interesting patterns have began to emerge, which still remain to be properly investigated.

3.3   Grammar Comparisons

Figure 9 displays three charts, depicting the number of rules, percentage of counter-examples and computation times of each of our grammars for D_n^3 with n ranging from 2 to 6 (where n denotes the number of abc triplets). Even though none of our proposed grammars is complete, we observe that as grammars get more abstract, the number of failing parses steadily declines. This however comes at the cost of rule size growth, which in turn is associated with an increase in computation times. What this practically means is that we are unable to continue testing more elaborate grammars or scale our results to higher orders of n (note that $\|D_n^3\|$ also has a very rapid rate of expansion[3]).

4 Discussion

To our knowledge, no other attempt has come as close to modelling D^3 with a 2-MCFG. We attribute this to the combination of a pragmatic approach with results from existing theoretical work. In this section, we present a collection of additional ideas, which we consider worthy of further exploration.

First-Match Policy and Relinking. Possibly the most intuitive way of checking whether a word w is part of D_n^3 is checking whether a pair of links occur that match a_i to b_i and b_i to c_i, $\forall\ i\ \in\ n$. We call this process of matching the *first-match policy*. The question arises whether a grammar can accomplish inserting a triplet of a, b, c, that would abide by the first-match policy. If that were the case, it would be relatively easy to generalize this ability by induction

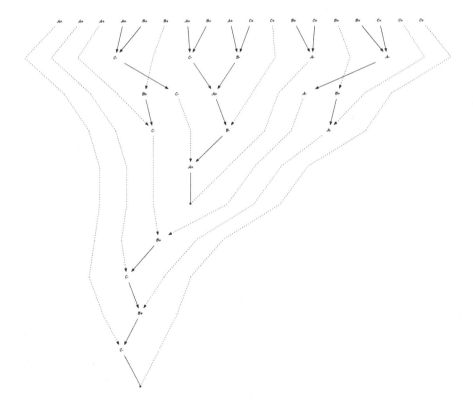

Fig. 8. Spider web of *"abaacbbacbabaccbcc"*

---

[3] https://oeis.org/A000108.

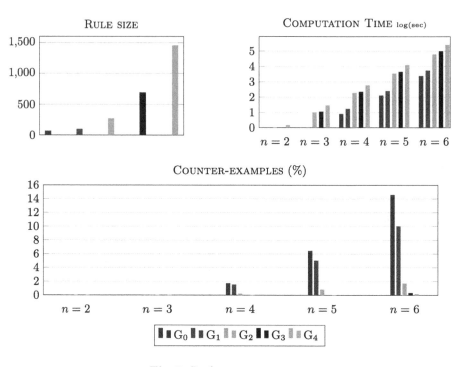

Fig. 9. Performance measures

to every $n \in \mathbb{N}$. Unfortunately, the answer is seemingly negative; the expressiveness provided by a 2-MCFG does not allow for the arbitrary insertions required. On a related note, being able to produce a word state $W(x, y)$ where $w = xy$ and x any possible prefix of w, gives no guarantee of being able to produce the same word with an extra triplet inserted due to the straddling property.

However, if rules existed that would allow for match-making and breaking, i.e. match *relinking*, an inserted symbol could be temporarily matched with what might be its first match-policy in a local scope, and then relink it to its correct match when merging two words together.

Growth Rules. Although $\mathcal{G}_2$ comes close to realizing the growth algorithm, not all of the growth rules presented in Fig. 5 can be translated into a 2-MCFG setting. In particular, there is no straight-forward way to render the growth rules which produce two edges (i.e. the six bottom-left rules); two neighbouring edges can in turn interact with their next neighbours in a recursive fashion, which can not be accommodated by our current tuple representation. We cannot, however, rule out the possibility of the growth rules being scalably translatable into m-MCFG rules without resorting to more expressive formalisms (e.g. context-sensitive rewriting systems). In fact, such a translation would be a guarantee of completeness.

Fig. 10. First-match policy for *"ababacbcabcc"*

Insights from Promotion. An interesting question is whether promotion can be handled by a 2-MCFG. If so, it could be worth looking into the properties of orbits, to test for instance if there are promotions within an orbit that can be easier to solve than others. Solving a single promotion and transducing the solution to all equivalent words could then be a guideline towards completeness.

5   Conclusion

We tried to accurately present the intricacies of D^3 and the difficulties that arise when attempting to model it as a 2-MCFL. We have developed and introduced some novel techniques and tools, which we believe can be of use even outside the problem's narrow domain. We have largely expanded on the existing tools to accommodate MIX-style languages and systems of meta-grammars in general.

Despite our best efforts, the question of whether D^3 can actually be encapsulated within a 2-MCFG still remains unanswered. Regardless, this problem has been very rewarding to pursue, and we hope to have intrigued the interested reader enough to further research the subject, use our code, or strive for a solution on her own.

Acknowledgements. We would like to thank Michael Moortgat for introducing us to the problem, providing insightful feedback and motivating us throughout the process, as well as Jurriaan Hage for suggesting the use of multi-dimensional Dyck languages in static program analysis.

References

1. Chatterjee, K., Choudhary, B., Pavlogiannis, A.: Optimal Dyck reachability for data-dependence and alias analysis. Proc. ACM Program. Lang. **2**(POPL), 30 (2017)
2. Gazdar, G.: Phrase structure grammar. In: Jacobson, P., Pullum, G.K. (eds.) The Nature of Syntactic Representation. Synthese Language Library (Texts and Studies in Linguistics and Philosophy), vol. 15, pp. 131–186. Springer, Dordrecht (1982). https://doi.org/10.1007/978-94-009-7707-5_5
3. Gazdar, G., Klein, E., Pullum, G.K., Sag, I.A.: Generalized Phrase Structure Grammar. Harvard University Press, Cambridge (1985)
4. Kanazawa, M., Salvati, S.: MIX is not a tree-adjoining language. In: Proceedings of the 50th Annual Meeting of the Association for Computational Linguistics: Long Papers-Volume 1. pp. 666–674. Association for Computational Linguistics (2012)
5. Kuperberg, G.: Spiders for rank 2 Lie algebras. Commun. Math. Phys. **180**(1), 109–151 (1996)

6. Lewis, H.R., Papadimitriou, C.H.: Elements of the Theory of Computation. Prentice Hall PTR, New Jersey (1997)
7. Ljunglöf, P.: Practical parsing of parallel multiple context-free grammars. In: Workshop on Tree Adjoining Grammars and Related Formalisms, p. 144 (2012)
8. Moortgat, M.: A note on multidimensional Dyck languages. In: Casadio, C., Coecke, B., Moortgat, M., Scott, P. (eds.) Categories and Types in Logic, Language, and Physics. LNCS, vol. 8222, pp. 279–296. Springer, Heidelberg (2014). https://doi.org/10.1007/978-3-642-54789-8_16
9. Nederhof, M.J., Shieber, S., Satta, G.: Partially ordered multiset context-free grammars and ID/LP parsing. Association for Computational Linguistics (2003)
10. Petersen, T.K., Pylyavskyy, P., Rhoades, B.: Promotion and cyclic sieving via webs. J. Algebr. Comb. **30**(1), 19–41 (2009)
11. Reape, M.: Getting things in order. Discontinuous Const. **6**, 209–253 (1996)
12. Reps, T.: Program analysis via graph reachability. Inf. Softw. Technol. **40**(11–12), 701–726 (1998)
13. Reps, T., Horwitz, S., Sagiv, M.: Precise interprocedural dataflow analysis via graph reachability. In: Proceedings of the 22nd ACM SIGPLAN-SIGACT symposium on Principles of programming languages, pp. 49–61. ACM (1995)
14. Salvati, S.: MIX is a 2-MCFL and the word problem in Z^2 is solved by a third-order collapsible pushdown automaton. J. Comput. Syst. Sci. **81**(7), 1252–1277 (2015)
15. Seki, H., Matsumura, T., Fujii, M., Kasami, T.: On multiple context-free grammars. Theor. Comput. Sci. **88**(2), 191–229 (1991)
16. Wijngaarden, A.: The generative power of two-level grammars. In: Loeckx, J. (ed.) ICALP 1974. LNCS, vol. 14, pp. 9–16. Springer, Heidelberg (1974). https://doi.org/10.1007/978-3-662-21545-6_1

Compositionality in Privative Adjectives: Extending Dual Content Semantics

Joshua Martin[(✉)]

Department of Linguistics, Harvard University, Cambridge, USA
joshuamartin@g.harvard.edu

Abstract. Privative adjectives such as *fake* have long posed problems to theories of adjectives in a compositional semantics. In this paper, I argue that a theory like Del Pinal's recent Dual Content semantics, which encodes lexical entries with both an extension-determining component and a conceptual component, is best equipped to account for privativity while maintaining compositionality, but requires some revisions to do so. I provide some novel evidence for this system regarding recursive privativity to justify a new denotation for *fake*, use this to derive the patterns in some recent experimental data, and introduce an extension to the system to account for privative behaviors of intersective adjectives.

Keywords: Privative adjectives · Material adjectives · Modification · Conceptual structure · Lexical semantics · Qualia · Dual Content semantics

1 The Puzzle of Privativity

Most formal semantic approaches to modification begin from the assumption that both nouns and adjectives denote sets. Though their syntactic behavior is distinct, the meaning assigned to both is the set of all individuals of which the relevant property holds true. We can represent a noun like *dog*, then, as the set of all dogs, or formally, $\lambda x \in D_e$. $\mathrm{DOG}(x)$. Similarly, an adjective like *brown* is taken to mean the set of all brown things, formally $\lambda x \in D_e$. $\mathrm{BROWN}(x)$. These expressions are of the same semantic type, and so the members of both sets are drawn from the domain of individual entities D_e. Because of this, to interpret the full noun phrase *brown dog*, we can simply intersect the sets denoted by its component bare noun and attributive modifier, outputting the set of all things both *brown* and *dog*, given as $\lambda x \in D_e$. $\mathrm{BROWN}(x) \wedge \mathrm{DOG}(x)$.

The class of adjectives for which this property is true are, straightforwardly, termed intersective. These adjectives, which include basic properties such as

In writing this paper, I am grateful to Kathryn Davidson and the other members of the Meaning & Modality Lab at Harvard for enlightening discussions, the audience at ESSLLI 2018 at the University of Sofía for their valuable questions, and especially the reviewers whose insightful and generously detailed comments have reshaped it so much for the better. As always, all remaining errors are my own.

© Springer-Verlag GmbH Germany, part of Springer Nature 2019
J. Sikos and E. Pacuit (Eds.): ESSLLI 2018, LNCS 11667, pp. 93–107, 2019.
https://doi.org/10.1007/978-3-662-59620-3_6

color and shape, license inferences across head nouns, i.e. an [Adj X] that is also a Y is an [Adj Y]. Their meanings are noun-independent: if we know that *Toby is a brown dog* and that *Toby is a carnivore*, we can infer that *Toby is a brown carnivore*. In this way, his color is not true relative to the noun that *brown* happens to modify in any given sentence; it is a stable property of the individual. Additionally, statements about a bare noun entail statements about a modified NP produced by the addition of an intersective adjective: if something is true of all *mammals*, then it is also true of all *brown mammals*.

If all adjectives behaved so cooperatively, the semantics of modification would be a short-lived field. Fortunately, this clean intersective picture breaks down rather quickly. Subsective adjectives, by contrast, have meanings relativized to the noun which they modify. The set denoted by the modified NP can be characterized as a subset of the set characterized by the noun, namely that subset for which the adjective holds in the context of the noun. These adjectives, most commonly evaluatives like *skillful* or *excellent*, can be used predicatively, but these sentences are most naturally interpreted as still relative to a specific category (e.g. *Toby is skilled* carries some context-dependent meaning about precisely which skills he holds, while *Toby is brown* does not). Statements about the bare noun do still entail statements about the modified NP. Crucially, subsective adjectives do *not* license inferences across a head nouns; here, an [Adj X] that is also a Y is not necessarily an [Adj Y].

Beyond subsectives, which are tricky enough in their own right, are the nonsubsectives. Nonsubsective adjectives come in two varieties. Plain, or modal, nonsubsectives, which include *alleged, potential*, and *doubtful*, do not specify a set-theoretic relationship between the modified NP and the head noun. Statements about the bare noun do not entail statements about the modified NP, and inferences across nouns are not licensed. It is not true that all properties that hold of *thieves* also hold of *alleged thieves*, and we cannot conclude, if Toby is a *dog* and also an *alleged cookie thief*, that he is somehow an *alleged dog*.

This paper is concerned with the second type of nonsubsective adjective, known as privative adjectives. Privativity, loosely defined, is the phenomenon in which the reference of an adjective-noun pair is entirely disjoint from the reference of the noun alone; that is, the intersection of the set of individuals denoted by the adjective-noun phrase and the set denoted by the bare noun is the empty set. Statements about the bare noun do not entail statements about the modified NP (in fact, some probably entail the inverse, e.g. *guns kill people* likely entails *fake guns do not kill people*, though this is not a generalizable property, since *guns are black* certainly does not entail *fake guns are not black*. We will return to this issue.), and inferences across nouns are not licensed. Commonly cited members of this class include *fake, counterfeit, mock*, and *former*.

It is both deliberate and crucial that I chose a definition of privativity, however, not as a property inherent to certain adjectives, but as an emergent phenomenon occurring with certain adjective-noun combinations. Adjectives which seem in almost all ways to be intersective can, in fact, behave privatively when in the right contexts, modulating their behavior with respect to the noun that they modify. The most prominent example is constitutive material adjectives, such as *stone, metal, plastic*, which behave intersectively with nouns that can be made of the relevant material and privatively with others. The *stone* in *stone table*, for example, satisfies all of the intersective criteria, while the *stone* in *stone lion* behaves privatively, a stone lion likely not being classified as strictly a lion. We will also see that even canonically 'privative' adjectives do not universally exhibit the same behavior.

The privative behavior of adjectives like *fake* and *stone* in some contexts poses a notable challenge for compositional theories of semantics to grapple with. Since no elements of the bare noun set end up in the modified NP set, it is difficult to evaluate what contribution the bare noun makes to the compositional process, and what kind of operation the privative adjective is performing over it. While the entailment pattern suggests that the adjective is in a sense cancelling or excluding the denotation of the noun, it is clearly insufficient to say that a phrase like *fake guns* denotes 'the set of things which are not guns', which would include not only what we understand fake guns to be but also all other non-gun entities in the world.

Some have argued that compositional semantics cannot capture this pattern, and pursued analyses in non-compositional, conceptual frameworks, such as Franks [3] and Lakoff [7]. Within the compositional camp, nonsubsective adjectives initially led Kamp [6] to abandon the simple set-denoting description of even intersective adjectives, raising all adjective types as a means of 'generalizing to the worst case'. More recently, Partee [9] defends the possibility of a compositional analysis without making this move, and suggests that what we have heretofore considered privative nonsubsective adjectives are in fact neither of those things, and are actually a type of subsective adjectives. Their privative behavior arises from the fact that they 'coerce' the noun into an expanded meaning to avoid vacuity, and then pick out a subset of that expanded set. She argues that this is necessary to deal with data like (1):

(1)   a.   A fake gun is not a gun.
      b.   Is that gun real or fake?

Under most analyses, there is an obvious tension between the acceptability of both of these sentences, such that (1a) can be true while *fake* can also predicated of *that gun* in (1b). Partee argues that this is due to *fake* coercing *gun* into an expanded meaning in (1b), while the unmodified instance of *gun* in (1a) retains is original, limited denotation to the exclusion of fake guns. An expanded denotation of *gun* that can include fake guns is also argued to be necessary for felicitous use of *real*, which would otherwise be vacuous and redundant if all members of the set denoted by *gun* were always real guns. She encodes this in the form of two constraints: HEAD PRIMACY, which ensures that modifiers are interpreted relative to heads in complex phrases, and NON-VACUITY, which ensures that predicates resulting from modification have non-empty extensions. NON-VACUITY first causes an expansion of the denotation of *gun* so that *fake gun* is non-empty, before HEAD PRIMACY forces the modifier *fake* to be interpreted relative to the new interpretation of *gun* when it selects for a subset of the noun's new expanded set.

Her other arguments for privatives-as-subsectives, based on distributional patterns in Polish NP-splitting (see [10]), will not be reviewed here, but our concern here should be that while she motivates coercion as a mechanism, she does not provide the kind of formal representation that would allow us to determine the precise outcome of the process in this, or any given, instance. How, precisely, is the denotation of a noun coerced into a larger meaning, and perhaps more crucially, what is the larger meaning that a noun like *gun* is coerced into by the presence of *fake*? If it were merely expanded to include the presence of non-gun objects, this would surely overgenerate to a class far larger than what we would naturally consider fake guns. Can NON-VACUITY tell us how to expand the set differently when the noun is *gun* versus when the noun is *diamond*? The abstract description of the way constraints interact to *induce* coercion is insightful, but the set-theoretic *outcome* of the coercive process (namely, the expansion caused by NON-VACUITY) is left abstract. Whether a coercive account can, in principle, provide the compositional system with the tools to derive the intuitively correct output set by operating over purely extensional denotations is unclear. This is the problem that I think the following system addresses more successfully.

2   Introducing Dual Content

Del Pinal's Dual Content semantics (see [2]) preserves a system of composition nearly identical to classical function application, with minimal modifications, by adopting the assumption that the default lexical entries for nouns are notably more complex than in prior systems. On this view, common nouns have a binary semantic structure consisting of their extensional meaning (E-structure) and their conceptual meaning (C-structure). E-structure is the atomic extension-determining component, of the form $\lambda x.\text{STONE}(x)$. C-structure does not determine the extension of a noun, but instead consists of 'representations of perceptual features, functional features and genealogical features related to [the noun]' ([2], p. 4). These take the form of a Pustejovsky-style (see [12]) qualia structure. While only E-structure determines the extension of the noun's denotation, C-structure, it is argued, is a necessary component of speaker's linguistic competence in their ability to correctly identify members of a kind and use the term dynamically and productively in different contexts; it might be considered an instruction manual for correct and useful application of the linguistic term.

This non-atomistic theory of linguistic meaning may seem like a radical departure from the denotations formal semanticists are used to working with, but objections to purely definitional denotations have a rich history in the philosophy of language and mind, beginning with Putnam [13], whose criticisms formed the basis for a number of holistic theories of linguistic meaning.[1] Del Pinal's adaptation of the framework takes Pustejovsky's insights about the structure of the

---

[1] Of course, a rich philosophical history also brings with it a deep and heated debate. Holism is not without its critics. Here, we focus on the consequences for privativity, ignoring both Putnam's arguments against atomism and others' arguments against dual content-style representations, but an interested reader is pointed to Fodor [4] for these objections, and to Bilgrami [1] for a holistic response.

lexicon and endeavors to apply them to privative adjectives as a prime example of the kinds of tricky compositional meaning and pragmatic influence on composition that motivated such theories. Motivating this shift is a desire to unify the mechanical power of formal semantic frameworks with a 'psychologically realistic account of lexical semantics' [2]. To him, this kind of lexical structure is an insight from philosophy and psychology that linguists cannot afford to ignore, and privative adjectives provide the ideal test case for demonstrating not only its plausibility but its necessity.

Towards that end, let us briefly see it in action. A sample Dual Content entry for *gun* is given below.

(2) $\llbracket \textbf{gun} \rrbracket =$
E-structure: $\lambda x.\textsc{Gun}(x)$
C-structure:
   Constitutive: $\lambda x.\textsc{Parts-Gun}(x)$
   Formal: $\lambda x.\textsc{Perceptual-Gun}(x)$
   Telic: $\lambda x.\textsc{Gen}\ e[\textsc{Shooting}(e) \land \textsc{Instrument}(e, x)]$
   Agentive: $\lambda x.\exists e_1[\textsc{Making}(e_1) \land \textsc{Goal}(e_1, \textsc{Gen}\ e[\textsc{Shooting}(e) \land \textsc{Instrument}(e, x)])]$

Not all of the precise semantics of the C-structure elements in (2) will come into play here; what is necessary is that the C-structure of *gun* encodes that it is composed of gun parts, has the perceptual form of a gun, is generally used in shooting events, and was made with the goal to be used in shooting events. Now that lexical entries are decomposed in this format, we also introduce operations which are able to access specific components of a lexical entry's meaning.

(3) *Qualia functions*: partial functions from the meaning of terms into their respective C-structure denotations, namely, constitutive, formal, telic, and agentive. The qualia functions are Q_C, Q_F, Q_T, Q_A. For example, using the denotation for *gun* in (2): $Q_C(\llbracket \textbf{gun} \rrbracket) = \lambda x.\textsc{Parts-Gun}(x)$

Adjectives can have a Dual Content structure as well. Intersective adjectives could theoretically be represented in a simpler manner, perhaps with only E-structure, such as $\llbracket \textbf{red} \rrbracket = \lambda D_C.\lambda x.D_C(x) \land \textsc{Red}(x)$, or even more simply as type $\langle e, t \rangle$ and composing with nouns using Predicate Modification. Privative adjectives, then, make use of these qualia functions, by operating over different elements of the noun's C-structure, preserving some in the resulting NP denotation and rejecting others. I will skip over much of Del Pinal's exposition and argument for how he arrives at this eventual lexical entry for *fake*, and simply present the final version. Here, D_C is the domain of 'ordered sets of the E-structure and C-structure of common Ns' ([2], p. 14).

(4) $[\![\mathbf{fake}]\!] =$
     E-structure: $\lambda D_C.[\lambda x.\neg Q_E(D_C)(X) \wedge \neg Q_A(D_C)(x) \wedge \exists e_2[\mathrm{MAKING}(e_2)\wedge$
          $\mathrm{GOAL}(e_2, Q_F(D_C)(x))]]$
     C-structure:
          CONSTITUTIVE: $\lambda D_C.Q_C(D_C)$
          FORMAL: $\lambda D_C.Q_F(D_C)$
          TELIC: $\lambda D_C.\neg Q_T(D_C)$
          AGENTIVE: $\lambda D_C.[\lambda x.\exists e_2[\mathrm{MAKING}(e_2) \wedge \mathrm{GOAL}(e_2, Q_F(D_C)(x))]]$

With this entry for *fake*, the resulting referent will have the constitutive and formal qualia of the noun, will not have the telic or agentive qualia, and will have a new agentive qualia suggesting that the referent was made with the goal of having the same formal qualia as the noun (i.e. being a convincing fake). In this formalism, the negation of a qualia function indicates that the function does not apply to that entity; perhaps a notation like $Q_T(D_C) = 0$ would be more natural, but I will preserve Del Pinal's notation here to avoid confusion with the original work. To compose this complex modifier with our complex noun in (2), we will need a more complex notion of function application, which Del Pinal (p. 20) provides:

(5) Dual Content Function Application (FA^{DC}):
     If α is a branching node, $\{\beta, \gamma\}$ is the set of α's daughters, and $[\![\beta]\!]_E$ is a function whose domain contains $[\![\gamma]\!]$, then $[\![\alpha]\!]_E = [\![\beta]\!]_E([\![\gamma]\!])$ and $[\![\alpha]\!]_C = \langle Q_C([\![\beta]\!])([\![\gamma]\!]), Q_F([\![\beta]\!])([\![\gamma]\!]), Q_T([\![\beta]\!])([\![\gamma]\!]), Q_A([\![\beta]\!])([\![\gamma]\!])\rangle.$

Per (5), the E-structure of a modifier takes in the E-structure of the noun as its argument, as does each C-structure take in its corresponding C-structure argument. Then by FA^{DC}, the result of applying $[\![\mathbf{fake}]\!]$ in (4) to $[\![\mathbf{gun}]\!]$ in (2) is:

(6) $[\![\mathbf{fake\ gun}]\!] =$
     E-structure:
     $\lambda x.\neg Q_E(D_C)([\![\mathbf{gun}]\!])(x) \wedge \neg Q_A([\![\mathbf{gun}]\!])(x) \wedge \exists e_2[\mathrm{MAKING}(e_2)\wedge$
     $\mathrm{GOAL}(e_2, Q_F([\![\mathbf{gun}]\!])(x))]$
     C-structure:
          CONSTITUTIVE: $Q_C([\![\mathbf{gun}]\!])$
          FORMAL: $Q_F([\![\mathbf{gun}]\!])$
          TELIC: $\neg Q_T([\![\mathbf{gun}]\!])$
          AGENTIVE: $\lambda x.\exists e_2[\mathrm{MAKING}(e_2) \wedge \mathrm{GOAL}(e_2, Q_F([\![\mathbf{gun}]\!])(x))]$

Thus, we get a class of entities which are not guns, do not have the origins of guns, and were made to appear as if they were guns, but do not have the purpose of guns (i.e. are generally not used in shooting events). In the next section, I will provide some additional, previously unreported evidence for a Dual Content-style system, and extend the system to account for additional privative types.

3 Extending the System

3.1 *Fake fake guns* and Recursive Privativity

Any system which does not encode a non-extension-determining part of a lexical entry, even if it invokes qualia structure, will struggle to account for recursive applications of privative adjectives. If there is only one atomic component of the lexical entry, a privative must negate it entirely. Thus, a secondary application, as in *fake fake gun*, will negate that negation, and thus return functionally the original entry for *gun*. Is this an adequate denotation for *fake fake gun*? That is, should the two *fakes* cancel each other out in this way? It is certainly true that a *fake fake gun* could be a *gun*. But it is certainly not true that every *gun* is a *fake fake gun*, and not every *fake fake gun* would need to be a *gun*. The recursive form likely involves some element of deception, such as a criminal designing a real gun to appear like a toy in order to sneak it by security, but could also be a faked set of toy guns, not making them guns themselves. Of course, we would not naturally call standard-issue military or hunting weapons *fake fake guns*.

Dual Content gets closer to the intended meaning. The original denotation for *fake* in (4) will not quite do the job, but this system is at least equipped to provide a denotation that will. To first show the problems with the existing formulation, taking the denotation of *fake gun* from (6) and feeding it into the function *fake* from (4) produces the following output:

(7) $[\![$**fake fake gun**$]\!]$ =
   E-structure: $\lambda x. \neg Q_E(D_C)([\![$**fake gun**$]\!])(x) \wedge \neg Q_A([\![$**fake gun**$]\!])(x) \wedge$
   $\qquad \exists e_3[\text{MAKING}(e_3) \wedge \text{GOAL}(e_3, Q_F([\![$**fake gun**$]\!])(x))]$
   C-structure:
   $\qquad$ CONSTITUTIVE: $Q_C([\![$**fake gun**$]\!])$
   $\qquad$ FORMAL: $Q_F([\![$**fake gun**$]\!])$
   $\qquad$ TELIC: $\neg Q_T([\![$**fake gun**$]\!])$
   $\qquad$ AGENTIVE: $\lambda x. \exists e_3[\text{MAKING}(e_3) \wedge \text{GOAL}(e_3, Q_F([\![$**fake gun**$]\!])(x))]$

Which, since the constitutive and formal qualia of *fake gun* are simply that of *gun*, can be simplified further to:

(8) $[\![$**fake fake gun**$]\!]$ =
   E-structure: $\lambda x. \neg Q_E(D_C)([\![$**fake gun**$]\!])(x) \wedge$
   $\qquad \neg \exists e_2[\text{MAKING}(e_2) \wedge \text{GOAL}(e_2, Q_F([\![$**gun**$]\!])(x))] \wedge$
   $\qquad \exists e_3[\text{MAKING}(e_3) \wedge \text{GOAL}(e_3, Q_F([\![$**gun**$]\!])(x))]$
   C-structure:
   $\qquad$ CONSTITUTIVE: $Q_C([\![$**gun**$]\!])$
   $\qquad$ FORMAL: $Q_F([\![$**gun**$]\!])$
   $\qquad$ TELIC: $\neg Q_T([\![$**fake gun**$]\!])$
   $\qquad$ AGENTIVE: $\lambda x. \exists e_3[\text{MAKING}(e_3) \wedge \text{GOAL}(e_3, Q_F([\![$**fake gun**$]\!])(x))]$

But this gives us a contradiction in the E-structure. The agentive quale of a *fake gun*, as per (6), specifies that there exists a 'making' event with the goal of creating something with the formal quale of a gun. Since *fake* passes through the formal quale of the original noun to its output, Q_F of *gun* and *fake gun* are identical, giving us the contradictory result in (8) that there both was and was not an event in which x was made to look like a gun.

To resolve this issue, I propose this modification to the denotation for *fake*:

(9)   Dual Content denotation for *fake*, Version 2:

$\llbracket$**fake**$\rrbracket$ =
E-structure: $\lambda D_C.[\lambda x.\neg Q_E(D_C)(x) \wedge \neg Q_A(D_C)(x) \wedge$
$\exists e_2[\text{MAKING}(e_2) \wedge \text{GOAL}(e_2, \text{PERCEPTUAL-}Q_T(D_C)(x)]$
C-structure:
FORMAL: $\lambda D_C.\text{PERCEPTUAL-}Q_T(D_C)$
TELIC: $\lambda D_C.\neg Q_T(D_C)$
AGENTIVE: $\lambda D_C.[\lambda x.\exists e_2[\text{MAKING}(e_2) \wedge$
$\text{GOAL}(e_2, \text{PERCEPTUAL-}Q_T(D_C)(x))]]]$

This formulation makes a number of revisions. First, it removes any specification for the constitutive quale, on the intuition that the *fake* version of something is actually unlikely to use the same parts, or at least all the same parts. Second, it changes the formal quale to say that, rather than perceptually resembling the original object, the output perceptually resembles something with the telic quale of the original object. For example, a *fake gun* looks like something that does what a *gun* does, and thus a *fake fake gun* looks like something with the same purpose as a *fake gun* - namely, to look like a gun rather than to act like a gun. Third, it adjusts the part of the E-structure and the agentive quale which make reference to the formal quale to incorporate the previous change.

The precise formalism used to represent this notion of PERCEPTUAL-Q_T could be revised. One might be tempted to say that there is an event of perceiving in which one perceives x as the thing which is being *faked*, but that would be too strong, since it would require that the fake is successful to ever be considered a fake. That would rule out situations where one might naturally want to describe something as a *poor fake gun*, for example, suggesting that its intended purpose was not achieved. I am not especially committed to whichever functional mechanism is used here; it seems like one will have to be innovated regardless, and I cannot avoid the stipulation of some additional machinery here, but its precise nature is less important at the moment than the insight.

To see (9) in action, we can apply it to *gun* and get the following:

(10) $\llbracket$**fake gun**$\rrbracket$ =
E-structure: $\lambda x.\neg Q_E(D_C)(\llbracket$**gun**$\rrbracket)(x) \wedge \neg Q_A(\llbracket$**gun**$\rrbracket)(x) \wedge$
$\exists e_2[\text{MAKING}(e_2) \wedge \text{GOAL}(e_2, \text{PERCEPTUAL-}Q_T(\llbracket$**gun**$\rrbracket)(x))]$
C-structure:
FORMAL: PERCEPTUAL-$Q_T(\llbracket$**gun**$\rrbracket)$

TELIC: $\neg Q_T([\![\mathbf{gun}]\!])$

AGENTIVE: $\lambda x.\exists e_2[\text{MAKING}(e_2) \wedge$
$\qquad\qquad \text{GOAL}(e_2, \text{PERCEPTUAL-}Q_T([\![\mathbf{gun}]\!])(x))]$

By (10), a *fake gun* is something which is not a gun, does not have the origin or telos of a gun, and was made to look like it has the same purpose as a gun, or can do the same thing as a gun. This change also allows the recursive meaning to compute correctly:

(11) $[\![\mathbf{fake\ fake\ gun}]\!] =$

    E-structure: $\lambda x.\neg Q_E(D_C)([\![\mathbf{fake\ gun}]\!])(x) \wedge$
$\qquad\qquad \neg\exists e_2[\text{MAKING}(e_2)\wedge\text{GOAL}(e_2, \text{PERCEPTUAL-}Q_T([\![\mathbf{gun}]\!])(x))]$
$\qquad\qquad \wedge\exists e_3[\text{MAKING}(e_3)\wedge$
$\qquad\qquad\qquad \text{GOAL}(e_3, \text{PERCEPTUAL-}Q_T([\![\mathbf{fake\ gun}]\!])(x))]$

    C-structure:

    FORMAL: $\text{PERCEPTUAL-}Q_T([\![\mathbf{fake\ gun}]\!])$

    TELIC: $\neg Q_T([\![\mathbf{fake\ gun}]\!])$

    AGENTIVE: $\lambda x.\exists e_3[\text{MAKING}(e_3) \wedge$
$\qquad\qquad \text{GOAL}(e_3, \text{PERCEPTUAL-}Q_T([\![\mathbf{fake\ gun}]\!])(x))]$

A *fake fake gun*, now, is an object which: is not a *fake gun*, was not made to look like a *gun*, was made to look like a *fake gun* (and does), and does not have the same purpose as a *fake gun*, i.e. to look like something which shoots. A complex denotation, to be sure, but to my intuitions an accurate one, as a *fake fake gun* should certainly be trying to look like something which is itself pretending to be able to shoot, whether or not the *fake fake* version can shoot. This denotation accurately captures the intuition that *fake fake gun* says nothing about whether its members are actually guns or not.

The importance of this telic agnosticism can be illustrated by a few different contexts in which we might use *fake fake gun*. Consider the customs example from before. A criminal might want to smuggle guns across the border, and to that end repaint some guns to look like toys, with the bright orange tips characteristic of *fake guns*, so that the inspecting officer believes them to be toys when in fact they are deadly weapons. Alternately, a toy company could design a toy which mimics another toy. Perhaps you have a lighter, or candy dispenser, that looks exactly like a toy gun. Then there is a first 'faking' where a toy is made to look like it is a gun, and a second 'faking' in which a different item is made to look like the faked gun. Both of these scenarios we might describe as *fake fake guns*, even though the former items very much function as guns, and the latter items very much do not. In these examples, we can see that what is important is that the *fake fake gun* is made to imitate a *fake gun*, and that whether it shoots or not can vary by context. This suggests a sort of 'one step up' degree of preservation of telicity - the *fake fake gun* can call on the telic information of the *fake gun*, but has undergone too much modification to either maintain or reject the telic information of the bare noun.

This problem demonstrates that DC handles complex and iterative compositions in a more natural way than non-DC systems, and elucidates an instance in which the simplicity of other systems lead them to fail to accurately capture the meaning of a term which should be straightforwardly compositional.

3.2   Patterns of Entailment

Dual Content also shows promise in covering some recent experimental data on the inference patterns of privative noun phrases. Pavlick and Callison-Burch [11] show that speakers do not treat all instances of privative adjectives as entailing a negation of the head noun; specifically, speakers do still infer entailments between the privative-modified NP and the bare noun in certain cases that would not be predicted by the traditional analysis of privativity, which would predict that statements about the privative-modified NP should in fact contradict, not entail, the same statement about the bare noun. Some adjective-noun combinations 'behave in the prototypically privative way' ([11], p. 117), e.g. *counterfeit money* contradicts rather than entails *money*. Others behave contrary to this prediction, e.g. a *fake ID* is judged to still be an *ID*, a *mythical beast* is judged to still be a *beast*, and a *mock debate* is judged to still be a *debate* (or more specifically, statements that are true of the full NP are judged to entail the same statements about the bare noun).

The account of this data in Dual Content could, in spirit, follow rather naturally from the denotations for different privatives. Specifically, since different privatives negate different aspects of the C-structure, it is reasonable that you would occasionally retain entailment in some contexts if what is negated is unvalued in the noun for that particular quale. It does not seem incompatible with DC, either, that speakers would assign different relative pragmatic weights to different qualia in the C-structure, even if they aren't entirely absent.

Mock debate was judged to entail *debate* at a much greater rate than *mock execution* entailed *execution*. *Debate* probably is highly specified for its form (i.e. that it involves exchanging arguments in an oral format, most likely) but either unspecified or underspecified for its telos – perhaps to convey to the public some range of arguments, but also perhaps for competitive glory, or to convince voters to support you, but none of these seem crucial to the activity. Since *mock*, analyzed similarly to *fake*, negates the telic quale but not the formal quale, entailment relations are likely to hold between the unmodified and modified NP in the case of *debate*. Contrast with *execution*, which is likely underspecified for its form (i.e. it may be a firing squad, gallows, guillotine, electric chair, lethal injection, or any number of less typical methods) but highly specified for its telos – it must involve ending someone's life. If *mock*, then, negates the telos but not the form, then a *mock execution* is very unlikely to hold the same entailment relations as a regular *execution*, and in fact will be considered an explicit contradiction with it.

At this point, one might object, and say that, despite any intuitive appeal, both the original denotations for privative adjectives like *fake* and our revision in (9) prevent this from happening, because they require that the privative-modified set is extensionally disjoint from the original noun set. While adjectives like *mock* and *counterfeit* do vary in which C-structural element they target, they are invariant in that they include the $\neg Q_E(D_C)$ element in their own E-structure. This is true, and if we take these experimental results to mean that a *fake ID* is an *ID*, then there is no way for the system as it exists to capture this behavior. However, I would hedge against this conclusion due to the particular experimental design used. Since they asked about the likelihood of statements about one class entailing the same statement about the other class, and not explicitly about categorization or set identity, we can argue that what is actually being tapped into here is C-structure, or as originally defined, the range of associated beliefs that competent speakers have about members of that class, which determine their general use in language but not their extension. Future experimental work should certainly investigate whether speakers have this postulated divergence in their conceptual judgments of object properties and their category-membership judgments of object term extensions, but as it is, the existing data seems to be a strong argument in favor of a system that can capture this variance between adjectives and their dynamic interaction with different nouns.

This type of analysis also predicts the strangeness, possibly vacuousness, of certain privative NPs, such as *counterfeit light* (referring not to an object like a lamp, but to the light itself). *Counterfeit* is analyzed as similar but not identical to *fake*, in that it negates the agentive quale of the noun and requires that the object was made with the goal of looking and behaving like the noun ([2], p. 16). Thus, since *light* probably lacks an agentive quale, being predominately an experiential phenomenon and capable of being produced from a number of different sources and reactions, *counterfeit light* would be a strangely redundant utterance as effectively nothing is being negated. Partee's Principle of Full Interpretation (see [9]) might apply here, inducing pragmatic infelicity where *counterfeit* makes no substantive contribution to the meaning.

3.3 Pseudo-Privative Predicates

Classical privative adjectives like *fake* are not the only instances of privativity in our adjectival typology. Other adjectives, most commonly used in intersective or predicative ways, sometimes behave privatively. Take the case of constitutive material adjectives like *stone*. They are often simply intersective, as in *stone door*, and can be predicated of individuals, as in *That door is stone*. But they also behave privatively, such as in *stone lion*, referring to a statue rather than a real lion. In this section, I will argue for an extension of DC to cover constitutive material adjectives, while retaining uniformity between their predicative and privative uses. Del Pinal briefly discusses constitutive material adjectives, but does not develop lexical entries nor show compositions for them. I propose that the lexical entry for *stone* is something like the following:

(12) $[\![\mathbf{stone}]\!] =$
    E-structure: $\lambda x. Q_C([\![\mathbf{stone}]\!])(x) \wedge Q_A([\![\mathbf{stone}]\!])(x)$
    C-structure:

CONSTITUTIVE: $\lambda x.\text{STONE}(x)$
AGENTIVE: $\lambda x.\text{EXCAVATED}(x) \lor \text{CARVED}(x)$

This lexical entry is, most notably, recursive. The E-structure accesses elements of the C-structure of the same term, such that the extension of *stone* is determined by the C-structure of its members. The C-structure of another term may not be extension-determining, but adjectives that target a certain kind of qualia structure (such as *constitutive* material adjectives) may take these qualia to be necessary to satisfy their extension. We have already seen examples of promoting C-structure elements to E-structure with Del Pinal's denotation for *typical*. This denotation for *stone* is type $\langle e, t \rangle$ and so will compose with its same-typed head noun *lion* through Predicate Modification.

(13) $[\![\mathbf{lion}]\!] =$
       E-structure: $\lambda x.\text{LION}(x)$
       C-structure:
           CONSTITUTIVE: $\lambda x.\text{SUBSTANCE-LION}(x)$
           FORMAL: $\lambda x.\text{PERCEPTUAL-LION}(x)$
           AGENTIVE: $\lambda x.\exists e_1[\text{BIOLOGICAL-BIRTH-LION}(E_1,X)]$

Note that *lion* is unspecified for the telic quale. Straightforward Predicate Modification with these denotations would lead to the following denotation for *stone lion*:

(14) $[\![\mathbf{stone\ lion}]\!] =$
       E-structure: $\lambda x.\text{LION}(x) \land Q_C([\![\mathbf{stone}]\!])(x) \land Q_A([\![\mathbf{stone}]\!])(x)$
       C-structure:
           CONSTITUTIVE: $\lambda x.\text{SUBSTANCE-LION}(x) \land \text{STONE}(x)$
           FORMAL: $\lambda x.\text{PERCEPTUAL-LION}(x)$
           AGENTIVE: $\lambda x.\exists e_1[\text{BIOLOGICAL-BIRTH-LION}(E_1,X)] \land (\text{EXCAVATED}(x) \lor \text{CARVED}(x))$

But this is problematic. A *stone lion* should not be a lion, nor should it be composed of biological lion parts, or the result of a lion birth, and these things are explicitly contradictory with the constitutive and agentive qualia of *stone* such that the resulting set is empty, a violation of Non-Vacuity. In fact, all that we want from *lion* is its formal quale, namely having the perceptual features or shape of a lion. To achieve this effect while preserving the predicative uses of *stone*, we will have to slightly modify our notion of Predicate Modification. To this end, I introduce the notion of *E-Precedence*. Its formal definition is given in (15), but the basic intuition is as such: whenever, in Predicate Modification, one element is specified for a part of qualia structure in its E-structure, and the other element is either unspecified for that quale or specified for it only in its C-structure, the element which has the quale in its E-structure will win out and its value for the given quale will override the C-structural value for the corresponding quale.

(15)   Predicate Modification with E-Precedence (PM^{EP}):
       If α is a branching node, $\{\beta, \gamma\}$ is the set of α's daughters, $[\![\beta]\!]$ and $[\![\gamma]\!]$
       are both in $D_{\langle e,t \rangle}$, and $[\![\beta]\!]_E = \lambda x. \forall I' \in I, Q_{I'}(x)$ where $I \subset \{C, F, T, A\}$,
       then $[\![\alpha]\!]_E = \lambda x. [\![\beta]\!]_E(x) \ \wedge \ \forall J' \in J, Q_{J'}([\![\gamma]\!])(x)$, where $J \cap I = \varnothing \ \wedge$
       $J \cup I = \{C, F, T, A\}$.

In (15), I and J stand for subsets of the set of qualia. This will produce an
output where the E-structure of the composed phrase includes the E-structure
of β and any elements of the C-structure of γ which are not specified in the
E-structure of β. Since the only element of C-structure that is specified in *lion*
and not in the E-structure of *stone* is the formal, applying PM^{EP} to *stone* and
lion gives us the following result:

(16) $[\![\text{stone lion}]\!]_E = \lambda x. [\![\text{stone}]\!]_E(x) \wedge Q_F([\![\text{lion}]\!])(x)$
       $= \lambda x. \text{STONE}(x) \wedge (\text{EXCAVATED}(x) \vee \text{CARVED}(x))$
       $\wedge \text{PERCEPTUAL-LION}(x)^2$

This seems to be a good match for our intuitions: an object made of stone,
produced through some kind of stonework, and with the perceptual form of a
lion. This approach, then, builds in some of the basic insight of an Optimality
Theory-style system for adjective composition, such as Oliver's Interpretation as
Optimization (see [8], itself built on ideas from Hogeweg [5]) – namely, the notion
that some features are more highly-ranked than others, which will not necessarily
always cause an override when they are compatible, but can do so – into the
existing functional composition system. It also allows a uniform denotation for
constitutive material adjectives like *stone* to capture their simple predicative and
complex privative behaviors. While the introduction of this additional condition
on the Predicate Modification rule is undesirable by parsimony, it does not seem
possible for any solution to the problem of pseudo-privativity to be achieved
without either allowing non-uniform lexical entries or some kind of modification
to FA or PM, and E-Precedence seems a rather natural one following from Dual
Content. This innovation will not change the analysis for true privatives like *fake*,
since they compose through FA rather than PM regardless. For non-privative
uses of intersective adjectives, we will simply see no clash in C-structure and
therefore no override, and PM^{EP} will behave identically to standard PM.
    As a final note for this section, Del Pinal observes that the word *literally*
can instruct the listener to relax their commitment to Non-Vacuity and accept
seemingly empty denotations in the case of constitutive material adjectives, but
not in the case of true privatives:

---

2 A reviewer points out that, if *stone* were to be E-structurally specified for its perceptual quale, this analysis would not work. My simple answer to this objection is
   that *stone* clearly is not - I don't think anyone would say that if a stone item were
   made to look even completely like something else, it would no longer count as stone
   in the way that it would not were it actually composed of some other material - and
   were it, then we would probably want to predict a contradiction here. If both β and
   γ are E-structurally specified for the same quale, then (15) would simply conjoin
   them and if they are contradictory, we get an infelicitous utterance.

(17)   a. Something unbelievable happened in a laboratory at Harvard. Scientists discovered a way of making, literally, stone lions.

        b. Something amazing happened in a laboratory at Harvard. Some engineer managed to make, literally, a fake gun.

The observation is that we are willing to imagine a hypothetical living lion composed of stone in the case of (17a), but that we cannot do any parallel operation for (17b). In the current analysis, we could explain this by saying that *literally* is a signal for the listener to ignore E-Precedence in their interpretation. This would affect the interpretation of constitutive material adjectives, which are composed using PM, but not of true privatives, which use FA, matching the observed pattern.

3.4   Conclusion: Returning to the Puzzle

Partee [9] lays out two problems for privativity: can it be accounted for in a compositional semantics, and can so-called privative nonsubsective adjectives be given a subsective denotation that explains their patterning with respect to NP splits in Polish? She argues with respect to the latter that privative adjectives coerce the noun into an expanded meaning, but the implementation of this coercion is left unspecified. I have argued that Del Pinal's Dual Content semantics offers the least theoretically costly account of privativity (adhering most closely to compositionality in the Fregean sense) while also showing the most explanatory power. By expanding the lexical entries for nouns to include an extension-determining and an associated non-extension-determining conceptual component, Dual Content allows a treatment of privative adjectives as, in a sense, subsective, since they pick out referents with some of the conceptual features of the bare noun while excluding others.

Cases of iterated privativity like *fake fake guns* posed a problem for the proposed denotation of *fake*. A revised denotation of the adjective allows Dual Content to accommodate these cases, and it still does so more successfully than its competitors, as extensional-only semantics seem unlikely to be able to capture this behavior. I also have suggested that Dual Content offers an explanation for some puzzling recent experimental data, and that a uniform analysis of each privative adjective can still account for their inconsistent application across head nouns, though additional empirical investigation of these patterns is needed.

Finally, I argued that introducing the notion of E-Precedence into Predicate Modification, such that the extension-determining components of a modifier can override the conceptual components of a noun when there is a clash, allows an extension of the Dual Content framework to account for the privative behaviors of constitutive material adjectives while preserving a uniform lexical entry for those adjectives when they are used predicatively and intersectively. This innovation integrates some of the insight of the Interpretation as Optimization theory into a functional semantics, while avoiding its downfalls.

Many questions about the adjective typology still remain, including an effective way to implement non-privative non-subsective adjectives. These questions,

I suspect, will be answered with tools that are perfectly compatible with, though do not depend on, Dual Content, such as degree semantics for adjectives involving time (e.g. *former*) and possible world semantics for adjectives involving possibility (e.g. *potential*). Open questions also remain about the extent of Dual Content, namely which linguistic elements need this kind of split structure and whether any other types of modification need to make reference to it. For now, I hope to have shown that a functional semantics can capture our intuitions about multiple types of privativity, and some of its less obvious behaviors, and more fully specify what expanded denotation privative adjectives coerce their nouns into. This approach highlights two essential properties of privativity: first, that it can in a sense be considered a type of subsectivity as identified by Partee, and second, that it is better analyzed as an emergent phenomenon that dynamically arises from certain adjective-noun combinations, rather than a typological description of certain adjectives, with the unifying factor of including some override of the extensional structure of the noun, either through the adjective's inherent denotation or through an operation like E-Precedence.

References

1. Bilgrami, A.: Why holism is harmless and necessary. Philos. Perspect. **12**, 105–126 (1998)
2. Del Pinal, G.: Dual content semantics, privative adjectives, and dynamic compositionality. Semant. Pragmat. **8**(7), 1–53 (2015)
3. Franks, B.: Sense generation: a "quasi-classical" approach to concepts and concept combination. Cognit. Sci. **19**, 441–505 (1995)
4. Fodor, J., Lepore, E.: Holism: A Shopper's Guide. Blackwell, Cambridge (1992)
5. Hogeweg, L.: Rich lexical representations and conflicting features. Int. Rev. Pragmat. **4**, 209–231 (2012)
6. Kamp, H.: Two theories about adjectives. In: Keenan, E.L. (ed.) Formal Semantics of Natural Language. Cambribge University Press, Cambribge (1975)
7. Lakoff, G., Johnson, M.: Metaphors We Live By. The University of Chicago Press, Chicago (1980)
8. Oliver, M.: Interpretation as optimization: constitutive material adjectives. Lingua **149**, 55–73 (2013)
9. Partee, B.: Formal semantics, lexical semantics, and compositionality: the puzzle of privative adjectives. Philologia **7**, 11–21 (2009)
10. Partee, B.: Privative adjectives: subsective plus coercion. In: Zimmerman, T., et al. (eds.) Presuppositions and Discourse: Essays Offered to Hans Kamp, pp. 273–285. Emerald Group Publishing, Bingley (2010)
11. Pavlick, E., Callison-Burch, C.: So-called non-subsective adjectives. In: Proceedings of the Fifth Joint Conference on Computational Semantics, pp. 114–119 (2016)
12. Pustejovsky, J.: The generative lexicon. Comput. Linguist. **17**(4), 409–441 (1991)
13. Putnam, H.: Is semantics possible? Metaphilosophy **1**(3), 187–201 (1970)

Definiteness with Bare Nouns in Shan

Mary Moroney[(✉)]

Cornell University, Ithaca, NY, USA
mrm366@cornell.edu
http://conf.ling.cornell.edu/mmoroney/about.html

Abstract. Shan, a Southwestern Tai language spoken in Myanmar, Thailand, and nearby countries, uses bare nouns to express both unique and anaphoric definiteness, a definiteness distinction identified by [15]. This novel data pattern from the author's fieldwork can be analyzed by adding an anaphoric type shifter, ι^x, to the available type shifting operations defined by [6] and [7]. Otherwise, the data from Shan fit well with predictions from a type shifting analysis as laid out by [8]. Additionally, this paper demonstrates that the consistency test [7] is not sufficient to determine what counts as a definite determiner for a language.

Keywords: Bare nouns · Definiteness · Type-shifting · Tai language

1 Introduction

[15] proposes that there are two types of definiteness expressed by German, corresponding to the contracted (*weak*) and non-contracted (*strong*) preposition + definite article combinations—e.g., *vom* ('by the', *weak*) and *von dem* ('by the', *strong*). In (1), the speaker and listener know that there is only one mayor in the context. Since the mayor is unique in the context, the weak definite article form, *vom* ('by the') is used and the strong form is infelicitous.

(1) WEAK VERSUS STRONG ARTICLES IN GERMAN ([15]: (42))
     Der Empfang wurde **vom**   / #**von dem**   Bürgermeister
     the reception was   by-the$_{weak}$ / by   the$_{strong}$ mayor
     eröffnet.
     opened
     'The reception was opened by the mayor.'

[15] claims that the split between strong and weak definite forms fits well with the types of definiteness described by [9], who grouped definiteness into four categories: immediate situation (current non-linguistic context), larger situation (broader non-linguistic context), anaphoric/familiar, and bridging (associative anaphora). [15] says that when a noun is unique in an immediate

Thanks to Nan San Hwam, Mai Hong, and Sai Loen Kham who provided the Shan data. Thanks also to Sarah Murray, Miloje Despic, the Cornell Semantics Group, and my reviewers for all their feedback. Any errors are my own.

J. Sikos and E. Pacuit (Eds.): ESSLLI 2018, LNCS 11667, pp. 108–123, 2019.
https://doi.org/10.1007/978-3-662-59620-3_7

situation or larger situation context, German uses the weak form of the definite article, and in anaphoric contexts it uses the strong form. For the bridging category, he discusses two types: producer-product and part-whole bridging, which I will call 'product-producer' and 'whole-part' bridging, respectively.[1] The strong form is used in product-producer situations and the weak form is used in whole-part bridging. In addition to the categories discussed by [9,15] adds that donkey anaphora uses the strong form of the definite article. Table 1 gives examples of these categories and the article form used for German. These will be discussed more in the following section.

Table 1. Types of definiteness described by [15], citing [9]

Type of definite use	Example	German
Unique in immediate situation	The desk (uttered in a room with exactly one desk)	weak
Unique in larger situation	The prime minister (uttered in the UK)	weak
Anaphoric	John bought a book and a magazine. The book was expensive	strong
Bridging: Product-producer	John bought a book today. The author is French	strong
Bridging: Whole-part	John was driving down the street. The steering wheel was cold	weak
Donkey anaphora	Every farmer who owns a donkey hits the donkey	strong

Section 2 discusses the unique and anaphoric types of definiteness identified by [15], primarily looking at data from Thai from [12]. Section 3 introduces data demonstrating the types of definiteness available for bare nouns in Shan, a language related to Thai.[2] Sect. 4 gives a type shifting analysis for Shan bare nouns using two types of ι type shifters and discusses some problems with using the consistency test to decide whether a word counts as a determiner for the Blocking Principle ([7]). Section 5 concludes.

---

[1] A review noted that what [15] calls 'part-whole' bridging would more correctly be called 'whole-part' bridging, and agreeing with their assessment, I will use that and 'product-producer' instead of 'producer-product' for the same reasons.

[2] Data for this paper comes from the author's fieldwork with the Shan language in Chiang Mai, Thailand from January 2018 to present, working primarily with a speaker from Keng Tawng City in Shan State, Myanmar, who has lived in Thailand for over 10 years. Data was collected using a variety of elicitation methods: story translation, stories based on storyboards, felicity judgments on grammatical sentences in specific contexts, following techniques described in [3].

2   Background on Uniqueness and Anaphoricity

2.1   Uniqueness Versus Familiarity/Anaphoricity

[15] claims that the weak definite article in German expresses *uniqueness*. This can be uniqueness in an immediate situation, as in (2), or in a larger or global context, described further below. In (2), there is only one glass cabinet in the immediate context, so the weak definite must be used.

(2)   GERMAN: UNIQUE IN IMMEDIATE SITUATION ([15]: (40))
Das Buch, das du   suchst, steht   **im**   / **#in dem**
the book that you look-for stands in-the$_{weak}$ / in   the$_{strong}$
Glasschrank.
glass-cabinet

'The book that you are looking for is in the glass-cabinet.'

The strong definite article expresses *familiarity/anaphoricity*. In (3), the first sentence introduces a writer and a politician into the discourse context. In the second sentence *von dem Politiker* ('from the politician') is used to refer back to the politician. The strong definite form must be used in this context.

(3)   GERMAN: ANAPHORA ([15]: (23))
Hans hat einen Schriftsteller und **einen Politiker** interviewt. Er hat
Hans has a   writer   and a   politician interviewed He has
#vom   / **von dem**   **Politiker** keine interessanten Antworten
from-the$_{weak}$ / from the$_{strong}$ politician no   interesting   answers
bekommen.
gotten

'Hans interviewed a writer and a politician. He didn't get any interesting answers from the politician.'

Looking at Mandarin and Thai, [11] and [12] show that these languages use bare nouns in the same places where German would use the weak definite article, and noun phrases combined with a classifier and demonstrative (N Cl Dem) where German would use the strong definite article. Examples (4) and (5) show the use of the bare noun in a unique situation in Mandarin and Thai, respectively.

(4)   MANDARIN: UNIQUE IN IMMEDIATE SIT. ([11]: (12b), citing [5]: 510)
Gou yao guo malu.
dog   want cross road
'The dog(s) want to cross the road.'

(5)   THAI: UNIQUE IN IMMEDIATE SITUATION ([12]: (2))
măa kamlaŋ hàw.
dog   PROG   bark
'The dog is barking.'

In (6) and (7), are the Mandarin and Thai examples using demonstratives to express familiarity/anaphoricity. In (6a), a boy and a girl are introduced into the discourse context. (6b) and (6c) use *na ge nasheng* ('the/that boy'), a noun with a classifier and demonstrative, to refer back to the boy. In Mandarin there is a contrast between the subject and object position. The classifier and demonstrative are optional in subject position, but not in object position, as shown in (6b) and (6c). [11] claims that this is because the Mandarin subject is a topic, which negates the need for an antecedent index.[3]

(6)   MANDARIN: NARRATIVE SEQUENCE (ANAPHORIC) ([11]: (16a,b,d))
  a. jiaoshi   li   zuo-zhe **yi ge nansheng** he **yi ge**
     classroom inside sit-PROG one CLF boy   and one CLF
     nüsheng,
     girl
     'There is a boy and a girl sitting in the classroom...'

  b. Wo zuotian   yudao #(**na ge**) **nansheng**
     I   yesterday meet   that CLF boy
     'I met the boy yesterday.'

  c. (**na ge**) **nansheng** kan-qi-lai you er-shi sui zuoyou.
     that CLF boy   look   have two-ten year or-so
     'The boy looks twenty-years-old or so.'

In the Thai example in (7), the first sentence introduces a student into the discourse context. In (7a), *nákrian khon nán* ('that boy'), which has a classifier and demonstrative combined with the noun, is used to refer to the boy. To get the anaphoric reading, the demonstrative is required even in subject position in Thai.

(7)   THAI: NARRATIVE SEQUENCE (ANAPHORIC) ([12]: (17))
  mîawaan phǒm cəə   kàp **nákrian khon nɨŋ**.
  yesterday 1ST   meet with student CLF   INDEF
  'Yesterday I met a student'

  a. **nákrian** #(**khon nán**) chalàat mâak.
     student CLF   that clever very
     'That student/that one was very clever.'

2.2   Associative Anaphora (Bridging)

[15] shows that in German, there is a split between whole-part and product-producer bridging in terms of definiteness marking: whole-part bridging uses the weak definite form and product-producer bridging uses the strong definite form. [11] and [12] show that Mandarin and Thai pattern with German, using the bare

---

[3] [15] also notes that in cases where a weak definite article can refer anaphorically, the role of the referent is a topic in that it is 'the one that the story is about' (47).

noun in whole-part examples (weak definiteness) and the classifier-demonstrative combination in product-producer examples (strong definiteness). In this section and the following one, only the Thai data is shown to conserve space.

In (8), *thábian* ('license') cannot be combined with a demonstrative. This parallels the use of the weak definite for whole-part bridging in German.

(8) THAI: WHOLE-PART BRIDGING ([12]: (11))

rót khan nán thùuk   tamrùat sàkàt   phrɔ́ʔ mâj.dâj tìt
car CLF that ADV.PAS police   intercept because NEG   attach
satikəə wáj thîi **thábian** (#**baj nán**).
sticker keep at license CLF   that

'The car was stopped by police because there was no sticker on the license.'

In (9), the producer *náktèɛŋklɔɔn* ('poet') must be combined with a demonstrative. This parallels German's use of the strong definite for product-producer bridging.

(9) THAI: PRODUCT-PRODUCER BRIDGING ([12]: (12))

ʔɔɔl khít   wâa   **klɔɔn** bòt nán prɔ́ʔ   mâak, mɛ̂ɛ-wâa kháw cà
Paul thinks COMP poem CLF that melodious very although 3P   IRR
mâj chɔ̂ɔp **náktèɛŋklɔɔn** #(**khon nán**).
NEG like   poet   CLF   that

'Paul thinks that poem is beautiful, though he doesn't really like the poet.'

2.3 Donkey Anaphora

In cases of donkey anaphora, [15] claims that German uses the strong article to refer to nouns introduced in the first part of the construction. Similarly, in Thai and Mandarin, a demonstrative is required in those positions [11,12]. For the Thai example in (10), using a bare noun to refer back to the buffalo gives the sentence a generic meaning 'Every farmer that has a buffalo hits buffalo'.

(10) THAI: DONKEY ANAPHORA ([12]: (23))

chaawnaa thúk khon thîi mii **khwaai tua nɨŋ**   tii **khwaai tua**
farmer   every CLF that have buffalo CLF INDEF hit buffalo CLF
nán
that

'Every farmer that has a buffalo hits [that buffalo].'

Table 2 summarizes the patterns of definiteness expression in German, Thai, and Mandarin. Examples of all the contexts described by [15] for all three languages can be found in the cited sources.

3   Aspects of Definiteness in Shan

3.1   Uniqueness Versus Familiarity/Anaphoricity

Just like Mandarin and Thai, Shan—a Southwestern Tai language spoken in Myanmar—uses the bare noun in unique situations, as shown in (11) and (12).[4] In (11), there is a single teacher in the context, so it must be referred to using a bare noun. In (12), world knowledge tells us that there is only one sun, so a bare noun is used to refer to the sun. The demonstrative is not felicitous in either case.

(11)   SHAN: UNIQUE IN IMMEDIATE SITUATION
    (Context: classroom with just one teacher)

    Náaŋ Ľɤn ʔàm tsaaŋ kwàa hǎa **khúsɔ̌n** (#kɔ̂   nân).
    Ms. Lun NEG able go   find teacher CL.PERSON that

    'Ms. Lun cannot find the teacher.'

(12)   SHAN: UNIQUE IN LARGER SITUATION

    kǎaŋwán (#hɔ̀j   nân) lǒŋ hɤ   sɔ̀ŋ.
    sun   CL.ROUND that very bright glitter

    'The sun is very bright.'
    (Speaker comment on the demonstrative: there is more than one sun)

    Unlike Mandarin and Thai, Shan can use the bare noun in anaphoric contexts such as a narrative sequence. In (13), the first sentence introduces a man into the discourse context.[5] In following sentences, the man can be referred back to either using a bare noun, *phu-tsáaj* ('man'), or using noun combined with a classifier and demonstrative (N Cl Dem), *phu-tsáaj kɔ̂ nân* ('that man').

Table 2. Expressions of definiteness in German, Thai, and Mandarin

Type of definite use	German ([15])	Thai ([12])	Mandarin ([11])
Immediate situation	weak	bare	bare
Larger situation	weak	bare	bare
Anaphoric	strong	dem.	dem.
Bridging: Product-producer	strong	dem.	dem.
Bridging: Whole-part	weak	bare	bare
Donkey anaphora	strong	dem.	dem.

---

[4] Glossing conventions: 1: first person, 3: third person, CL: classifier, COMP: complementizer, INF: informal, IMPF: imperfect, IRR: irrealis, NEG: negation, PASS: passive, PL: plural, PRF: perfect, PRT: particle, SG: singular.

[5] Another individual, the store owner, is introduced in the story because otherwise the most natural thing to use anaphorically is the pronoun *mán-tsáaj* ('him').

(13) SHAN: NARRATIVE SEQUENCE (ANAPHORA)
phu-tsáaj kô nɯɯŋ kwàa ti hâan khăaj măa tàa sǔɯ măa
person-man CL.PERSON one go   at store sell dog for buy dog
ʔɔn tǒ   nɯŋ păn luk jíŋ mán-tsáaj... **phu-tsáaj**
small CL.ANIMAL one give child girl 3-man   person-man
(**kô**   **nân**) khɯ́n tɔ̀p   waa,
CL.PERSON that back respond that
'A man went to a dog store to buy a puppy for his daughter... The/that
man replied,'

In (14), the first sentence introduces a notebook and cup of water into the
discourse context. The second sentence refers back to each of them using a bare
noun. Here the anaphoric nouns are in object position, but this position does
not require that a demonstrative be used. In this way, Shan is different from
Mandarin or Thai. The demonstrative is allowed, but it sounds awkward to use
a demonstrative for both the water cup and notebook in the second sentence.

(14) SHAN: NARRATIVE SEQUENCE (ANAPHORA)
pâp măaj lɛ **kɔ́k nâm** jù   wâj nɤ̆ phɤ̆n. khaa qăw **kɔ́k nâm**
book note   and cup water IMPF stay on desk 1.SG take cup water
(**nân**) he   sàj/saǔɯ **pâp**   (**nân**).
that   spill in   CL.BOOK that
'There is a notebook and a cup of water on the desk. I spilled the/that
cup of water onto the/that notebook.'

In (15) is another example with anaphora in the object position. Again, the
classifier and determiner are optional.

(15) SHAN: NARRATIVE SEQUENCE (ANAPHORA)
tsɔn   tǒ   nɯɯŋ máa mí   nɤ̆ tonmâj ʔăn   mí   hímtsăm
squirrel CL.ANIMAL one   come exist on tree   COMP exist near
hɤ́n   háw nân sě   mɤnâj phǒn lóm haaŋ   hâaj nàa lɛ hét
house 1PL that and today rain wind appearance bad very and do
haj kìŋ-mâj   jàat tók njăa tě phât njáa **tsɔn** (**tǒ**
cause branch-tree break fall almost IRR hit   meet squirrel CL.ANIMAL
nân) páa jâw
that with PRF
'A squirrel was on the tree near my house. Then one day, a bad storm
caused a branch to break and fall almost hitting the squirrel.'

3.2 Associative Anaphora (Bridging)

Mandarin, Thai, and German use the weak/bare form of the nominal in whole-
part bridging and the strong/demonstrative form in product-producer bridging. Shan, however, does not require the strong/demonstrative form in product-
producer bridging. A bare noun can be used in both situations. (16) shows that
a bare noun is possible and likely required for whole-part bridging in Shan.

(16)   SHAN: WHOLE-PART BRIDGING

khúsŏn kwàa tsú **hýntŕk** lăŋ   nân sĕ tòj   **pháktŭ** hôŋ
teacher go   to building CL.BUILDING that and knocked door   call
tsaw hýn.
owner building

'The teacher approaches that building and knocked on the door to call
the owner.'

Whole-part bridging constructions in Shan often have the 'whole' as part of
the word for the 'part'. It is not always clear whether it simply anaphoric with the
'whole' possessing the part or involves bridging to a real noun compound. (17)
shows an example of this where *naasɤ pâplik* ('book cover') contains the word
pâplik ('book').[6] While it is possible for the noun to appear with a demonstrative,
the demonstrative is referring to the book rather than the cover. It does not seem
possible for the bridged noun to be combined with a demonstrative.

(17)   SHAN: WHOLE-PART BRIDGING

mɛ́w wĕnkjɔ́k sàj nɤ̆ **pâplik** ʔăn   mí   nɤ̆ phɤ̆n măa kwàa tsɔ́m
cat jump in on book COMP exist on table dog go   follow
theŋ. thŭŋ ti   hét haj   **naasɤ pâplik** (**nân**) kokòmkolĭn kwàa
again until COMP do cause cover book   that   dirty   go
seŋ.
completely

'The cat jumped onto the book that was on the table. The dog followed
again which made the book cover/cover of that book completely dirty.'

(18) shows that a demonstrative is not necessary for product-producer bridging either. The 'producer', *kóntɛmlik* ('author') can be bare or a N Cl Dem
phrase, *kóntɛmlik kɔ̂ nân* ('that author'). From the classifier we can tell that
this demonstrative-classifier combination is attached to 'author' not 'book'.

(18)   SHAN: PRODUCT-PRODUCER BRIDGING

mɤwáa   khú   ʔàan **pâplik púɯn** táj. khúsŏn pĕn ʔɔ́jkɔ̂ kăn
yesterday teacher read book   history Tai teacher be friend together
táŋ **kóntɛmlik** (**kɔ̂   nân**).
with author   CL.PERSON that

'Yesterday, the teacher read a Tai (Shan) history book. The teacher is
friends with the/that author.'

---

[6] As [1] note for Akan and [2] notes for Bulu, a possessive construction seems required
to do whole-part bridging in those languages. The contrast between cases like (16),
where a bare noun seems to be an option, and cases like (17), where a possessive
construction seems to be required suggest this possibility should be investigated in
more depth for Shan.

3.3   Donkey Anaphora

German, Thai, and Mandarin use the strong/demonstrative form of the nominal
to refer anaphorically to a nominal in donkey anaphora. Unlike the Thai and
Mandarin, Shan does not use a demonstrative or strong definite article in this
situation. In (19), when 'cat' (*mɛ́w*) is referred to anaphorically, a bare noun is
used. It is not felicitous to use a N Cl Dem phrase because that forces a singular
reading, which sounds awkward in this sort of generic sentence.

(19)   SHAN: DONKEY ANAPHORA
       mǎa ku   tŏ   nâj pɔ́   hăn **mɛ́w** nǎj tɛ̌ lɯp   lám
       dog every CL.ANIMAL this if/when see cat   then will follow chase
       mɛ́w (*tŏ   **nân**) tàasè.
       cat   CL.ANIMAL that always
       'Every dog, if it sees a cat/cats will always chase the cat(s).'

If we wanted to use a N Cl Dem construction in this sort of example, a
structure like (20) would be possible, but, again, the demonstrative and classifier
are not necessary. The difference between these two examples is that in (19) it
is dogs being quantified over, leaving 'cat' as unspecified for plurality and thus,
perhaps, awkward with a singular anaphor. In (20), *tŏ lǎj* ('which one') quantifies
over individual cats making it compatible with a singular anaphor.

(20)   SHAN: DONKEY ANAPHORA
       mǎa nâj hăn **mɛ́w tŏ**   **lǎj**   kɔ tɛ̌ lɯp   **mɛ́w** (tŏ
       dog this see cat   CL.ANIMAL which PRT will follow cat   CL.ANIMAL
       nân) tàasè.
       that always
       'Dogs, whichever cat they see they will always chase the/that cat'

In donkey anaphora constructions, my consultants resisted introducing 'one
cat' in the if-clause portion. If 'only' is used, it is okay, and in that case it
is necessary to use a demonstrative to refer back to it, as in (21). This raises
interesting questions about what kind of antecedent a bare noun sets up and
why donkey anaphora might function differently from other anaphora contexts.

(21)   SHAN: DONKEY ANAPHORA
       mǎa ku   tŏ   nâj pɔ́   hăn **mɛ́w tŏ**   **nɯŋ** kój
       dog every CL.ANIMAL this if/when see cat   CL.ANIMAL one   only
       kɔ tɛ̌ lɯp   lám   **mɛ́w** *(tŏ   **nân**) tàasè.
       PRT will follow chase cat   CL.ANIMAL that always
       'Every dog, if it sees only one cat, will always chase that cat.'

Table 3 summarizes the different expressions of definiteness found in German,
Thai, Mandarin, and Shan. This section has investigated the expression of definiteness in Shan in specific contexts that have shown different patterns of expression
across languages. Shan allows for the bare noun to be used in all of the contexts
described by [15]. Even contexts like anaphora and product-producer bridging

Table 3. Expressions of definiteness in German, Thai, Mandarin, and Shan

Type of definite use	German ([15])	Thai ([12])	Mandarin ([11])	Shan
Immediate situation	weak	bare	bare	bare (11)
Larger situation	weak	bare	bare	bare (12)
Anaphoric	strong	dem.	dem.	**bare** (13)–(14)
Bridging: Product-producer	strong	dem.	dem.	**bare** (18)
Bridging: Whole-part	weak	bare	bare	bare (17)
Donkey anaphora	strong	dem.	dem.	**bare** (19)–(20)

allow for bare nouns where Thai and Mandarin do not. For contexts were the noun is unique in a situation or with whole-part bridging, a classifier-demonstrative construction cannot appear with the noun, just like in Thai and Mandarin.

4   Analysis: Type-Shifting

4.1   A Type Shifting Analysis of Bare Arguments

Shan is similar to many other languages in the region in that nouns combined with numerals do not mark plurality, but do require a classifier for count nouns, as in (22)–(23), and require a measure word for mass nouns, as in (24)–(25).

(22) măa nuɯŋ *(tŏ)
      dog one CL.ANIMAL
      'one dog'

(23) măa săam *(tŏ)
      dog three CL.ANIMAL
      'three dogs'

(24) nâm nuɯŋ *(kɔ́k)
      water one cup
      'one cup of water'

(25) nâm săam *(kɔ́k)
      water three cup
      'three cups of water'

Languages that use classifiers in this way typically allow bare nouns to function as arguments. In English, only plurals and mass nouns can appear as bare arguments, as shown in (26)–(27).

(26)   Dinosaurs are extinct.

(27)   Gold is valuable.

To account for this, Chierchia [6] proposes a Neo-Carlsonian approach following [4], to claim that bare plurals in English have a mass-like denotation, of type $\langle e, t \rangle$, which can type shift, allowing them to function as arguments. The relevant type-shifting operators described by [6] and [7], are defined below:

(28)   TYPE SHIFTING OPERATORS ([7]):
       a. $\cap$: $\lambda P \, \lambda s \, \iota x[P_s(x)]$
       b. $\cup$: $\lambda k_{\langle s,e \rangle} \, \lambda x[x \leq k_s]$
       c. ι: $\lambda P \, \iota x[P_s(x)]$
       d. $\exists$: $\lambda P \, \lambda Q \, \exists x[P_s(x) \wedge Q_s(x)]$

[7] and [6] use the Blocking Principle, defined in (29), to identify what type shifting is available in what language. If a language has an overt determiner form of a type shifter—e.g., *the* in English is said to correspond to ι—then covert type shifting using that operator is unavailable.

(29)   BLOCKING PRINCIPLE [7]: For any type shifting operation ϕ and any X: $*\phi(X)$ if there is a determiner D such that for any set X in its domain, $D(X) = \phi(X)$.

In a language where there are no determiners, you would expect all type shifting operations to be available, but according to [7], $\exists$-type shifting does not occur in these languages. [7], revising [6], proposes that the type shifting operators follow a hierarchy, where kind-forming $\cap$ and entity forming ι must be ruled out before $\exists$ becomes available. This rule is defined in (30). The justification is that using $\cap$ or ι is a less drastic change because it does not introduce quantificational force. [7] claims that bare nouns are equally allowed to form kinds or entities, so they must be ranked equally.

(30)   MEANING PRESERVATION: $\{\cap, \iota\} > \exists$

While $\exists$ is only available as a type shifting operator when the others are ruled out. However, there is a narrow scope existential reading available for bare plurals in English that can be seen in negated sentences, as in (31). [6] analyzes this reading as coming from Derived Kind Predication, defined in (32), which fixes type mismatches between the verb and the argument.

(31)   John didn't see machines.   ([6]: (49d′))

(32)   *Derived Kind Predication (DKP)*   ([6]: 364)
       If P applies to objects and k denotes a kind, then $P(k) = \exists x[^{\cup}k(x) \wedge P(x)]$

In Shan, existential readings are available for bare nouns, as in (33). When negation is present, only the narrow scope existential reading is available (34), just like with English bare plurals. There are no restrictions on the existential reading based on the availability of pseudo incorporation as [8] propose for Teotitlán del Valle Zapotec (TdVZ) bare singulars, thus making Shan bare nouns appear similar to TdVZ bare plurals or mass nouns.

(33)   SHAN: EXISTENTIAL
       kǎw   hǎn mǎa
       1.SG(INF) see dog
       'I see a dog/dogs.'

(34)   SHAN: NARROW-SCOPE EXISTENTIAL

       a. kǎw   ʔàm laj hǎn mǎa   b. kǎw   ʔàm laj hǎn nâm
          1.SG(INF) NEG PRT see dog   1.SG(INF) NEG PRT see water
          'I didn't see dogs.' $[\neg > \exists]$ 'I didn't see water.' $[\neg > \exists]$

In addition to the existential and definite interpretations of Shan bare nouns discussed in this and previous sections, kind interpretations are also available for bare nouns, as in (35)–(36).

(35) SHAN: KIND PREDICATE WITH COUNT NOUN
      kóp mɔtwáaj   hăaj kwàa   jâw.
      frog CL.ANIMAL this disappear disappear go PRF

      'Frogs are extinct.'

(36) SHAN: GENERIC/KIND PREDICATE WITH MASS NOUN
      năm mí   lɤ́ŋ   năm wâj ku   ti
      water exist abundant plenty stay every place

      'Water is abundant everywhere.'

(37) SHAN: GENERIC/KIND PREDICATE WITH COUNT NOUN
      kóp mí lɤ́ŋ   năm wâj ku   ti
      frog exist abundant plenty stay every place

      'Frogs are abundant everywhere.'

According to [8], TdVZ mass nouns are kind-denoting (type $\langle s, e \rangle$) and bare plurals are sets containing pluralities (type $\langle e, t \rangle$). Following [6] and [7,8] summarize the available interpretations of bare mass nouns and bare plurals in languages without articles, claiming they can have a kind reading, a narrow scope existential reading, and a definite reading. This is summarized in Table 4.

Table 4. Expectations for bare mass and plural nouns in languages lacking definite articles [8]: Table 2, page 8

	Bare mass noun	Bare plural
(a) Kind-level reading	Expected N is kind-denoting	Expected ∩ type shift
(b) Narrow scope existential reading	Expected N denotation + DKP	Expected ∩ type shift + DKP
(c) Definite reading	Expected ∪ + ι type shift	Expected ι type shift
(d) Wide scope existential reading (∃ GQ interpretation)	Unexpected ∃ outranked by ι	Unexpected ∃ outranked by ι

This appears to be consistent with what is found in Shan. Table 5 identifies the examples discussed in this paper that demonstrate that Shan follows the predictions laid out by [8].[7]

---

[7] [6] would predict that a wide scope existential reading in cases where a kind and definite reading is blocked. This has not been tested yet.

Table 5. Interpretations available for Shan bare mass nouns and bare count nouns

	Bare mass noun	Bare noun
(A) Kind-level reading	Available (36)	Available (35), (37)
(B) Narrow scope existential reading	Available (34b)	Available (34a)
(C) Definite reading	Available (14)	Available (13)(13)
(D) Wide scope existential reading ($\exists$ GQ interpretation)	Unavailable (see (34b))	Unavailable (see (34a))

[6] claims that bare nouns in article-less languages without number marking, like Shan, obligatorily have a kind/mass-noun-like denotation of type $\langle s, e \rangle$. However, [8] allows for this type of nouns to undergo type shifting using $\cup$ so they can then type-shift using ι to get a definite reading separate from the kind reading. For now, I will assume this, following [8], but this topic should be considered in future work.

The predictions so nicely summarized by [8] do not distinguish between unique and anaphoric definiteness. This aspect of the analysis will be taken up in the following subsection.

4.2   Two Type of Definiteness

[11] follows [15] in claiming the existence of two types of definiteness. In trying to account for the obligatory use of the demonstrative in some definite environments in Mandarin, [11] defines the unique and anaphoric definites as in (38), where (38a) is the type shifting operation ι and (38b) is the denotation of the demonstrative in Mandarin.[8] [12] claims that since English expresses both unique and anaphoric definites using *the*, *the* is ambiguous for the unique and anaphoric definite meaning.

(38)   a.   UNIQUE DEFINITE ARTICLE:
  $[\![\iota]\!] = \lambda s_r . \lambda P_{\langle e, \langle s, t \rangle \rangle} . : \exists ! x [P(x)(s_r)] . \iota x P(x)(s_r)$
       b.   ANAPHORIC DEFINITE ARTICLE: ι^x
  $[\![\iota^x]\!] = \lambda s_r . \lambda P_{\langle e, \langle s, t \rangle \rangle} . \lambda Q_{\langle e, t \rangle} . : \exists ! x [P(x)(s_r) \wedge Q(x)] . \iota x P(x)(s_r)$

It is clear from the data that the Shan demonstrative does not fulfill the roll of anaphoric definite determiner since it is not obligatory in all anaphoric contexts as in Thai. I propose, instead that Shan has a null anaphoric type shifter ι^x in addition to the ι type shifter. Given that ι^x has more presuppositions,

---

[8] This definition differs from [15] in that the index is defined as a property rather than an individual, but I will not be concerned with this distinction for this analysis.

Maximize presupposition from [10] would predict that when ι and ι^x are in competition, ι^x should win out assuming that there is an indexical property available. However, since for a given situation and NP, ι must be true anywhere that ι^x is true, it is not possible to distinguish them.

This analysis raises the question: Why does the Shan demonstrative not count as a determiner for the purposes of the Blocking Principle, but the Thai and Mandarin ones do? We might expect the Shan demonstrative to pattern differently from the Mandarin and Thai demonstratives in terms of the Consistency test. [7] uses the Consistency test from [13] to distinguish between demonstratives and true definites. For demonstratives, you can introduce two identical N Cl Dem phrases with contradictory predicates, and there is no contradiction. For definite determiners, doing this would create a contradiction. According to this test, Shan has a demonstrative, not a definite determiner, as shown in (39). However, the Thai demonstrative also passes this test, as in (40).[9]

(39)   SHAN: CONSISTENCY TEST
       (Context: I am holding a white cup and a black cup.)
       kɔ́k hòj   **nâj** pĕn sĭ   khǎaw. **kɔ́k hòj**   **nâj** pĕn sĭ
       cup CL.ROUND this be   color white   cup CL.ROUND this be   color
       lǎm.
       black
       'This cup is white. This cup is black.'

(40)   THAI: CONSISTENCY TEST ([12], citing [14])
       dèk khon nán nɔɔn yùu tɛ̀ɛ **dèk khon nán** mâi.dâi nɔɔn yùu.
       child CLF   that sleep IMPF but child CLF   that NEG   sleep IMPF

       'That child is sleeping but that child is not sleeping.' (cf. #the)

According to a native Thai speaker, (40) sounds contradictory out of the blue, but fine with deixis. This test does not seem sufficient to distinguish between what counts as a definite for the Blocking Principle. This is not that surprising since the consistency test relies on deixis, which is not something that comes into play in anaphoric uses of demonstratives.

I would argue that the Shan bare noun/demonstrative contrast parallels the English *the*/demonstrative contrast. The difference comes from the fact that the bare noun in Shan can denote a broader range of things, which might lead to more disambiguation using the demonstrative. We would then expect the use of the demonstrative in Shan to convey some special meaning beyond ι in the same way the English demonstrative can.

5   Conclusion

Shan can use a bare noun to express both unique and anaphoric definiteness. In fact, the bare noun in Shan behaves much like the English article *the*. Though

---

[9] Mandarin passes the consistency test too, but the data is not included here to conserve space.

languages like Thai and Mandarin are similar to Shan in lacking overt definite articles and plural morphology, Shan does not pattern together with these two languages in that its demonstrative does not function as the primary marking of anaphoric definiteness. The pattern in Shan is likely to be found in other languages without articles, like Japanese and Russian.

This paper has also shown that the Consistency test does not seem able to distinguish what words count as determiners, so an avenue for future work would be to find more ways to identify 'true determiners'. In Mandarin and Thai, the demonstrative counts as the determiner denoting ι^x, so the demonstrative is obligatory in expressing this meaning. I argue that in Shan the demonstrative does not count as a determiner ι^x, so a bare noun can type shift using ι^x. It seems, then, that the anaphoric definite, ι^x, could be included as one of the available type shifting operations. This work in conjunction with the work by [12,15], and [11] brings up the connection between form and meaning. In Mandarin, Thai, and German there seems to be a connection between the obligatory use of a strong determiner/demonstrative and the need for an anaphoric index in the meaning. In Shan, that connection is unidirectional: if there is a demonstrative there must be an anaphoric index, but the lack of a demonstrative does not mean there is no anaphora involved.

Additionally, this paper has shown that the Shan data matches well with the predictions of the type shifting analysis from [6,7], and particularly [8]. Kind, definite, and narrow scope existential readings are all available for bare nouns in Shan. As a next step, an interesting line of research would be to see how a type shifting analysis of this sort affects how nouns can combine with other nominal elements, such as quantifiers, classifiers, or demonstratives.

References

1. Arkoh, R., Matthewson, L.: A familiar definite article in Akan. Lingua **123**, 1–30 (2013)
2. Barlew, J.: Salience, uniqueness, and the definite determiner-tè in Bulu. In: Snider, T., D'Antonio, S., Weigand, M. (eds.) Semantics and Linguistic Theory, vol. 24, pp. 619–639. LSA and CLC Publications, Ithaca (2014)
3. Bochnak, M.R., Matthewson, L.: Methodologies in Semantic Fieldwork. Oxford University Press, Oxford (2015)
4. Carlson, G.: A unified analysis of the English bare plural. Linguist. Philos. **1**(3), 413–457 (1977)
5. Cheng, L.L.-S., Sybesma, R.: Bare and not-so-bare nouns and the structure of NP. Linguist. Inq. **30**, 509–542 (1999)
6. Chierchia, G.: Reference to kinds across language. Nat. Lang. Semant. **6**, 339–405 (1998)
7. Dayal, V.: Number marking and (in)definiteness in kind terms. Linguist. Philos. **27**, 393–450 (2004)
8. Deal, A.R., Nee, J.: Bare nouns, number, and definiteness in Teotitlán del Valle Zapotec. In: Proceedings of Sinn und Bedeutung, vol. 21 (2017)
9. Hawkins, J.A.: Definiteness and Indefiniteness: A Study in Reference and Grammaticality Prediction. Croom Helm, London (1978)

10. Heim, I.: Artikel und Definitheit. In: von Stechow, A., Wunderlich, D. (eds.) Semantik: Ein internationales Handbuch der zeitgenösischen Forschung. de Gruyter, Berlin (1991)
11. Peter, J.: Articulated definiteness without articles. Linguist. Inq. **49**, 501–536 (2018)
12. Peter, J.: Two kinds of definites in numeral classifier languages. In: D'Antonio, S., Moroney, M., Little, C.R. (eds.) Semantics and Linguistic Theory (SALT), vol. 25, pp. 103–124. LSA and CLC Publications (2015)
13. Löbner, S.: Definites. J. Semant. **4**, 279–326 (1985)
14. Nattaya, P.: Classifiers and determiner-less languages: the case of Thai. Ph.D. dissertation, University of Toronto (2010)
15. Florian, S.: Two types of definites in natural language. Ph.D. dissertation, University of Massachusetts Amherst (2009)

The Challenge of Metafictional Anaphora

Merel Semeijn(✉)

University of Groningen, Groningen, The Netherlands
m.semeijn@rug.nl

Abstract. I argue that pronominal anaphora across mixed parafic-
tional/metafictional discourse (e.g. *In* The Lord of the Rings, *Frodo$_i$*
goes through an immense mental struggle. He$_i$ is an intriguing fic-
tional character!) poses a problem for current dynamic approaches to
fiction. I evaluate different possible solutions in a workspace account
based on a descriptivist approach, Maier's psychologistic DRT, Zalta's
logic of abstract objects and Recanati's dot-object analysis of fictional
characters.

Keywords: Metafictional statement · Parafictional statement ·
Workspace account · Anaphora · Abstract objects · Dot-objects

1 The Problem of the Wrong Kind of Object

Semanticists of fiction typically distinguish between (at least) three different
kinds of statements that contain fictional names (e.g. 'Frodo'): In Recanati's [22]
terminology, 'fictional', 'parafictional' and 'metafictional' statements. Fictional
statements are statements taken directly from a fictional work (e.g. (1) from *The
Lord of the Rings*). Parafictional statements are statements about the content
of a fictional work (e.g. (2) or (3) as found in a discussion on *The Lord of the
Rings*) that can be either 'explicit' (2) or 'implicit' (3), depending on whether the
'In fiction x,'-prefix is overt or covert. Metafictional statements are statements
about fictional entities *as fictional entities* (e.g. (4)):

(1) Frodo had a very trying time that afternoon.

(2) In *The Lord of the Rings*, Frodo lives in the Shire.

(3) Frodo lives in the Shire.

(4) Frodo was invented by Tolkien.

Any uniform semantic treatment of fictional names across these different types
of statements runs into a variation of 'the problem of the wrong kind of object'[1].
If we adopt a realist approach and assume that the name 'Frodo' refers to an
abstract object (e.g. Zalta [28,29] or Inwagen [11]) we run into difficulties with

---

[1] I adopt this term from Klauk [13] although he uses it to refer to only the realist
variant of the problem.

J. Sikos and E. Pacuit (Eds.): ESSLLI 2018, LNCS 11667, pp. 124–143, 2019.
https://doi.org/10.1007/978-3-662-59620-3_8

the interpretation of (2) and (3); Abstract objects are not the kind of things that can live in certain regions. On the other hand, if we adopt an antirealist approach and take the name 'Frodo' to refer to a flesh and blood individual in a set of counterfactual or pretense worlds (e.g. Lewis, [15], Walton [27] or Maier [18]), we run into difficulties with the interpretation of (4); Flesh and blood individuals are not the kind of things that can be invented.

An intuitive way out of these problems is to assume, following Kripke [14] and Currie [2], that fictional names are ambiguous; In statements such as (1), (2) and (3), 'Frodo' refers to a flesh and blood individual and in statements such as (4), 'Frodo' refers to an abstract object. However, consider the following discourse:

(5)   In *The Lord of the Rings*, Frodo$_i$ goes through an immense mental struggle to save his$_i$ friends. Ah yes, he$_i$ is an intriguing fictional character!

The metafictional statement in (5) contains a pronoun 'he' that is anaphorically dependent on the name 'Frodo' introduced in the preceding parafictional statement. Standardly, we take this to mean – contra Kripke and Currie – that the two terms co-refer. Such pronominal anaphoric links can potentially occur across all possible types of mixed discourse with fictional, parafictional and metafictional statements (See appendix) though some possibilities may be unlikely to actually occur. The current challenge for semanticists of fiction is thus to account for the use of pronominal anaphora across acceptable types of mixed discourse (rather than only being able to account for statements in isolation) while avoiding the problem of the wrong kind of object in its different variants.

In this paper, I discuss a dynamic semantic approach to modelling fictional and parafictional statements: 'The workspace account'[2] (Sect. 2) and argue that pronominal anaphora across mixed parafictional/metafictional discourse as in (5) (henceforth 'metafictional anaphora') poses a problem of accessibility for the workspace account (Sect. 3.1) that generalizes to other current dynamic semantic approaches to fiction (Sect. 3.2). I explore and evaluate four different possible solutions based on a descriptivist analysis of pronouns (Sect. 4.1), Maier's psychologistic DRT [18] (Sect. 4.2) Zalta's abstract object theory [28,29] (Sect. 4.3) and Recanati's dot-object analysis of fictional characters [22] (Sect. 4.4).

2   Introducing the Workspace Account

In Stalnaker's [25] widely adopted pragmatic framework, assertions are modelled as proposals to update the 'common ground' (i.e. the set of mutually presupposed propositions between speaker and addressee). The workspace account takes as its starting point Matravers' [20] theory of fiction. Following Friend [7], Matravers argues against the consensus view of fiction interpretation (e.g. Currie [2] or Walton [27]) that links fiction to the cognitive attitude of

---

[2] For details, see Semeijn [24].

imagination (i.e. regular assertions are mandates to *believe* and fictional statements are mandates to *imagine*). Matravers argues that there is no special cognitive attitude of imagination involved in our engagement with fiction. In fact, our primary engagement with both fictional and non-fictional narratives involves the same cognitive processes. Likewise, in the workspace account, fictional statements and regular assertions are modelled as proposals to update the same *temporary* common ground: the workspace. What differentiates non-fiction from fiction is whether, at the end of the possibly multi-sentence discourse, 'assertive' or 'fictive closure' is performed; Whether the content of the updated workspace is added to the common ground directly (for non-fiction) or as parafictional information (for fiction) under the relevant 'In fiction F'-operator. This operator receives a Lewisian [15] interpretation such that 'In fiction F, ϕ' is true iff ϕ is true at every world where F is told as known fact rather than fiction. In case a fictional statement is made and the common ground already contains parafictional information about the relevant fictional narrative (e.g. because there was a break in the story telling), 'fictive opening' is first performed (i.e. the content embedded under the fiction operator is placed in the workspace before updating).

I present a simplified representation of assertive closure of the assertion *Trump is the president of the U.S.* and of fictive closure of (1) from *The Lord of the Rings* (*LOTR*). To represent the workspace and common ground, I use the box notation of DRT (Discourse Representation Theory) developed by Kamp [12] in which NP's in a discourse are mapped to 'discourse referents' placed under several conditions in so-called DRS's (Discourse Representation Structures):

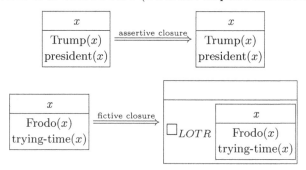

Whether I am reading a fictional narrative (e.g. *The Lord of the Rings*) or a non-fictional narrative (e.g. some article in *The Times*), I update the workspace with the content of the narrative. As soon as I stop entertaining the propositions of a non-fictional narrative I perform assertive closure: I stop updating the workspace (with e.g. *Trump is the president of the U.S.*), and instead update the common ground with this information (i.e. the result is that it is common ground that Trump is the president of the U.S.). As soon as I stop entertaining the propositions of a fictional narrative I perform fictive closure: I stop updating the workspace with fictional statements (e.g. *Frodo had a very trying time on a particular afternoon*) and instead update the common ground with *parafictional*

information based on the content of the workspace (i.e. the result is that it is common ground that in *The Lord of the Rings*, Frodo had a very trying time on a particular afternoon).[3]

Parafictional statements such as (2) and (3) are modelled as assertions about the content of a particular novel that exists in the actual world. Hence parafictional statements trigger assertive closure (i.e. the parafictional content of the workspace is added directly to the common ground):

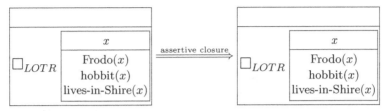

Thus, there are two ways to update the common ground with parafictional information; You either engage in fictional discourse (e.g. reading *The Lord of the Rings*) or you engage in parafictional discourse (e.g. engaging in a conversation about the content of *The Lord of the Rings*). What differs in these two cases is the content of the workspace (i.e. whether you entertained propositions about hobbits or parafictional propositions about the content of a particular novel).

3   The Challenge of Metafictional Anaphora

Metafictional statements pose a problem for the workspace account. More specifically, pronominal anaphora across mixed parafictional/metafictional discourse leads to a problem of accessibility. In this Sect. 1 introduce the problem of metafictional anaphora (Sect. 3.1) and show how it generalizes to other dynamic approaches (Sect. 3.2).

3.1   Metafictional Anaphora

Reconsider the central example of metafictional anaphora:

(5)   In *The Lord of the Rings*, Frodo$_i$ goes through an immense mental struggle to save his$_i$ friends. Ah yes, he$_i$ is an intriguing fictional character![4]

---

[3] As an anonymous reviewer noted, fiction about non-fictional objects (e.g. historical fiction) poses a challenge for this analysis. When we for instance read *"That's a fine death!" said Napoleon as he gazed at Bolkonski* in *War and Peace*, we update the workspace with information about Napoleon. If we then perform fictive closure as in the above DRS's, the discourse referent for 'Napoleon' will end up embedded. Arguably, *War and Peace* is about the *real* Napoleon – for which there probably already is a discourse referent in the main box – rather than about a fictional counterpart. Further research will have to produce an additional mechanism that links embedded discourse referents for non-fictional objects to their real counterparts.

Both sentences in (5) (a parafictional statement about *The Lord of the Rings* and a metafictional statement about Frodo) are assertions and hence trigger assertive closure on a workspace updated with (5). After updating with the parafictional statement in (5), the workspace looks as follows:

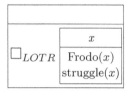

As Lewis [15] argued, metafictional statements are not covertly embedded under 'In fiction *F*'-operators. For instance, the metafictional statement in (5) does not express that in *The Lord of the Rings*, Frodo is an intriguing fictional character. In other words, there is no implicit 'In fiction *F*'-operator in the second half of (5), so the workspace is updated as follows:

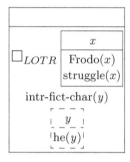

The pronoun 'he' in the metafictional statement in (5) triggers a presupposition that there is a masculine entity (denoted by the dashed box) (Cf. van der Sandt [23]) and we update with the information that this masculine entity is an intriguing fictional character.

The pronoun 'he' is anaphoric on the name 'Frodo' introduced in the preceding *parafictional* statement. Normally, we represent this by equating their discourse referents (i.e. resolving the presupposition and replacing all occurrences of y with x) so that the resulting update of the metafictional statement in (5) is 'intr-fict-char(x)'. However, following standard DRT-rules, x is not accessible outside of the 'In *The Lord of the Rings*'-operator and hence the presupposition remains unresolved. It is therefore unclear how we can update the workspace with (and thus interpret) metafictional statements involving metafictional anaphora such as in (5).

3.2 Other Dynamic Approaches

The problem of accessibility introduced by metafictional anaphora generalizes to other current dynamic approaches to fiction that involve some type of embedding

---
[4] For simplicity, I henceforth omit the anaphoric links of the possessive 'his'.

or separation of the content and discourse referents of the fictional narrative.[5] For instance, in Stokke's [26] and Eckardt's [3] 'unofficial common ground accounts' fictional and parafictional statements update a separated 'unofficial common ground' related to the relevant fiction and metafictional statements update the 'official common ground' related to actual states of affairs. The 'complete' common ground after updating with (5) will thus look as follows:

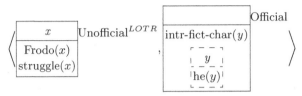

Although it is formally possible to make the discourse referents of fictional entities accessible outside of the unofficial common grounds where they were introduced, this is inconsistent with the motivation for having separate unofficial common grounds for fiction. Namely, that the content of fictional narratives is somehow quarantined from information about actual states of affairs. Moreover, in Stokke's framework unofficial common grounds are – unlike the official common ground – essentially temporary (i.e. they exist for the purpose and duration of the (para)fictional discourse). Making these temporary discourse referents accessible outside of the unofficial common grounds would lead to even more difficulties. Hence, in the unofficial common ground accounts, the presupposition triggered by the metafictional anaphora in (5) cannot straightforwardly be resolved because the discourse referent for 'Frodo' is not accessible outside of the unofficial common ground.

In the rest of this paper I focus on how to solve the problem of metafictional anaphora in the workspace account but the solutions under discussion should be extendable to the other dynamic approaches.

4   A Comparison of Different Solutions

In this Sect. 1 describe and evaluate four different strategies to meet the described challenge of accessibility posed by metafictional anaphora. Especially, I will discuss to what extent they avoid the problem of the wrong kind of object. First, one can adopt a descriptivist approach and account for metafictional anaphora in non-dynamic terms (Sect. 4.1). Alternatively, staying in the dynamic semantic framework, one can either adjust the accessibility relations (Sect. 4.2) or accommodate a new discourse referent that is accessible through standard accessibility relations. Such a discourse referent can be understood as an abstract object (Sect. 4.3) or as a dot-object (Sect. 4.4).

---

[5] With the possible exception of Maier's cognitive framework (See Sect. 4.2).

4.1   A Descriptivist Approach: A Description of Frodo

A possible solution to the described challenge in a traditional semantics framework is a descriptivist approach to anaphora (e.g. Evans [5], Elbourne [4] or Heim [10]). This analysis was originally proposed as a solution to the accessibility problem posed by donkey anaphora. Consider the following donkey sentence:

(6)   If Sarah owns a donkey, she beats it.

Intuitively, the pronoun 'it' does not refer to a particular individual donkey but is bounded by 'a donkey'. However, it is outside of the syntactic scope of 'a donkey' and hence inaccessible. On a descriptivist analysis, the anaphoric pronoun 'it' functions like, or 'goes proxy for', the definite description 'the donkey' retrieved from the preceding clause. In Elbourne's D-type account, this is because NPs at the level of syntax undergo phonetic deletion (are not pronounced at the surface level) when in the environment of an identical NP (e.g. *My shirt is the same as his*). Similarly, (6) is in fact equivalent to (7):

(7)   If Sarah owns a donkey, she beats the donkey.

This analysis evades the problem of the unbindable pronoun by replacing it with a definite description.

When we apply this strategy to (5), the pronoun 'he' is also analysed as going proxy for a definite description retrieved from the previous clause. However, (5) cannot be the result of simple phonetic deletion of an identical NP. If it were, (5) would be equivalent to something like (8):

(8)   In *The Lord of the Rings*, Frodo goes through an immense mental struggle to save his friends. Ah yes, the person named Frodo in *The Lord of the Rings* that goes through an immense mental struggle to save his friends, is an intriguing fictional character!

This gives us an incorrect analysis of (5): A flesh and blood person cannot be a fictional character. In other words, from the parafictional statement we retrieve the wrong kind of definite description and hence run into the antirealist variant of the problem of the wrong kind of object.

To get the correct interpretation what is required is a *metafictional* description such as 'the *character* named Frodo in *The Lord of the Rings*' so that (5) becomes equivalent to (9):

(9)   In *The Lord of the Rings*, Frodo goes through an immense mental struggle to save his friends. Ah yes, the character named Frodo in *The Lord of the Rings* is an intriguing fictional character!

Although, (9) gives an acceptable analysis of what is expressed by (5) it is unclear how to compositionally obtain such a meta-description of Frodo from the preceding clause. Moreover, even if we assume that we can accommodate such a definite description for metafictional anaphora, this solution does not extend to other types of mixed discourse such as pronominal anaphora across mixed metafictional/parafictional discourse (e.g. *Frodo$_i$ is an intriguing fictional character. Ah yes, in* The Lord of the Rings *he$_i$ goes through an immense mental* *struggle to save his friends!*) which would require accommodation of yet another type of definite description to avoid the *realist* version of the problem of the wrong kind of object.

Hence, a descriptivist approach does not (as yet) adequately account for metafictional anaphora; Simple phonetic deletion provides the wrong kind of definite descriptions and hence we need an account of how to accommodate the right kind of definite descriptions.

4.2 Psychologistic DRT: Adjusting Accessibility Relations

Another possible solution to the problem of metafictional anaphora (that sticks to a dynamic approach) is to adjust the accessibility relations so that the discourse referent for 'Frodo' *is* accessible.

Maier [18] adopts this strategy in his psychologistic DRT framework in which the context that is updated by statements is an agent's mental state. The agent's mental state is represented as a set of DRS's that are linked to cognitive attitudes such as belief (BEL) and imagination (IMG). Assuming that the agent updates with (5) after having engaged with *The Lord of the Rings* and having previously imagined Frodo, the agent's mental state after updating with (5) looks as follows:

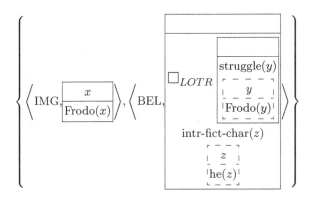

The parafictional and metafictional statement function similarly in that both update the belief-box and trigger presuppositions that need to take as their discourse referent x for 'Frodo'.[6]

Although this dynamic semantic framework – like the workspace account and the unofficial common ground accounts – involves separating the discourse referents of fictional entities, there is no accessibility problem because Maier assumes, contra usual practice, that attitudes can be referentially dependent on attitudes other than belief.[7] He gives the example of someone who wants to buy a new smartphone in a few years and imagines it to have a flexible

---

[6] Here I follow Geurts [9] in analysing proper names as triggering presuppositions.

[7] Maier does allow that there may turn out to be some structural constraints on specific cross-attitudinal dependencies.

transparent screen. This is a desire dependent imagination. Similarly, doxastic attitudes can be referentially dependent on imagination; When engaging in *The Lord of the Rings* I imagine the existence of an entity named Frodo and when engaging in parafictional or metafictional discourse such as (5) I believe that according to *The Lord of the Rings* this entity went through a mental struggle and that the entity is an intriguing fictional character. Hence the presuppositions triggered by the metafictional anaphora in (5) can be resolved; The content of the imagination-box is accessible.

Extending this strategy to the workspace account would amount to changing the accessibility relations relative to the content embedded under the 'In fiction F'-operator so that the presupposition triggered by the pronoun 'he' in the metafictional statement in (5) can be resolved:

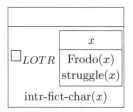

However, although there may be independent reasons to make the content in the imagination-box accessible in Maier's cognitive framework, there are none for doing this with the 'In fiction F'-operator, making this an *ad hoc* move. More importantly, there are already substantial theoretical costs in Maier's cognitive framework to making the content of the imagination-box accessible (i.e. a highly complex semantic system) but such a move with the 'In fiction F'-operator would amount to a drastic change of the basic semantics of DRT (in which any type of embedding entails inaccessibility). Such a radical departure of standard DRT semantics is *prima facie* undesirable.

In fact, the accessibility problem of metafictional anaphora may reappear in the psychologistic DRT framework as well. Because Maier assumes parafictional and metafictional statements are referentially dependent on existential imagination (induced by fictional statements) he only considers discourse in which the interpretation of parafictional and metafictional statements comes *after* the interpretation of fictional statements. However, fictional names can also be introduced in parafictional of metafictional statements. Suppose (5) featured the first occurrence of the name 'Frodo'. In this case the fictional name 'Frodo' introduced in the parafictional statement cannot be referentially dependent on a previous act of imagination. In Maier's framework interpretation of such parafictional discourse would have to involve either accommodation of a kind of contentless or minimal imagination during the parafictional discourse (e.g. imagining that there is a person named Frodo) or involve local accommodation of a discourse referent in the belief-box. The latter strategy would result in the following DRS:

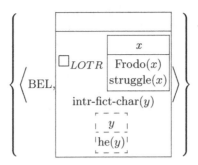

Here the discourse referent for 'Frodo' is embedded in the 'In *The Lord of the Rings*'-operator rather than the imagination-box. Making this discourse referent accessible will lead to the aforementioned problems of changing the accessibility relations relative to the 'In fiction *F*'-operator.

In any case, a solution based on changing the accessibility relations runs into the antirealist variant of the problem of the wrong kind of object. The imagined entity (in terms of the psychologistic DRT framework) or the entity that *The Lord of the Rings* is about (in terms of the workspace account) is a flesh and blood hobbit. If we make the discourse referent for this object accessible to the metafictional statement in (5), the metafictional statement will be about this flesh and blood individual. However, a flesh and blood hobbit cannot be an intriguing fictional character.

Hence, a solution based on Maier's psychologistic DRT does not adequately solve the problem of metafictional anaphora. It cannot straightforwardly be extended to a workspace account and runs into the antirealist version of the problem of the wrong kind of object.

4.3 Abstract Object Theory: Frodo the Abstract Object

An alternative strategy is to claim that fictional names in parafictional and metafictional statements refer to an object with a discourse referent that is accessible in the main box:

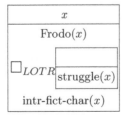

The above DRS formally resembles a DRS you would expect for parafictional discourse about a non-fictional character (e.g. *In* War and Peace, *Napoleon examines the Pratzen Heights*) where embedded parafictional information takes the discourse referent of an actually existing individual from the main box. However, contrary to Napoleon, Frodo never really existed. One way to account for this is to follow Zalta's application of his logic of abstract objects [28, 29] to fiction and claim that parafictional and metafictional statements are about abstract objects (e.g. Frodo the fictional character) that really exist.

Prima facie, this strategy seems to run head first into the realist version of the problem of the wrong kind of object; An abstract object cannot go through an immense mental struggle. To avoid this problem Zalta distinguishes between two types of objects and two modes of predication.[8] x is an 'ordinary object' ('O!(x)') if it is, or could have been, concrete (e.g. a chair). x is an 'abstract object' ('A!(x)') just in case it could not be concrete (e.g. the empty set). An ordinary object like a chair can 'exemplify' being red, i.e. it has the property of redness in the standard sense. Zalta denotes this using standard predicate logic notation: 'red(c)'. An abstract object can 'encode' a property which means it has this property as one of its constitutive characteristics. For instance, the empty set encodes the property of having no members. This is denoted with the argument to the left of the predicate: '$(\varnothing)$memberless'. Ordinary objects do not encode properties but abstract objects do exemplify properties (e.g. the empty set exemplifies being well-discussed: 'well-disc$(\varnothing)$').

Our fiction telling practices involve two types of abstract objects: 'stories' and 'fictional characters'. A story (e.g. *The Lord of the Rings*) is an abstract object that encodes the content of a narrative; It encodes 'vacuous' or 'propositional' properties of the form 'being such that P is true', where P is a proposition that is true in the story. A fictional character is an abstract object that is *native* to a story (e.g. Frodo or the One Ring, but not Napoleon). Contrary to common practice, Zalta draws a strong distinction between the analysis of explicit and implicit parafictional statements (e.g. respectively (2) and (3)). This is because Zalta is a realist about fictional characters (i.e. they exist as abstract objects) and hence we can talk about them as we do about ordinary objects (i.e. without an 'In fiction F'-operator or some type of pretense). A statement such as (3) is thus actually not 'implicit' in the sense that it is covertly embedded. Rather, it is a plain statement about what properties a certain abstract object encodes: '(f)lives-in-Shire'. Explicit parafictional statements (e.g. (2)) on the other hand do contain an 'In fiction F'-operator. They are statements about specific encoding and exemplifying relations between stories and characters. For instance (2) expresses that *The Lord of the Rings* encodes the vacuous property of being such that Frodo exemplifies living in the Shire: '$\square_{LOTR}$lives-in-Shire(f)'. Zalta proves a theorem in his theory:

(I) $\forall x \forall s(\text{Native}(x, s) \rightarrow \forall F(xF \equiv \square_s Fx))$

according to which, if some character x is native to some story s, implicit and explicit parafictional statements about x (in s) (e.g. (2) and (3)) necessarily follow from one another. Metafictional statements are statements about what properties fictional characters *exemplify*. For instance, the metafictional statement in (5) expresses that Frodo exemplifies the property of being an intriguing fictional character: 'intr-fict-char(f)'.

---

[8] This distinction originally comes from Mally [19].

Incorporating Zalta's analysis of parafictional statements into the workspace account suggests a modification of the fictive closure operation. Because of the strong distinction drawn between implicit and explicit parafictional statements, fictive closure can in theory involve two different kinds of updates of the common ground. An update with explicit parafictional statement and, if the relevant fictional characters are native to the relevant story, also with an implicit parafictional statement.[9] The following is a representation of fictive closure* of (1) that includes updates of the common ground with both types of statements:

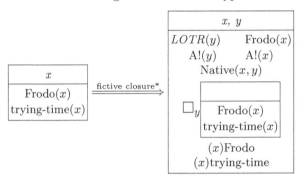

As soon as I stop reading *The Lord of the Rings*, I update the common ground with discourse referents for the newly introduced abstract objects (e.g. the story *The Lord of the Rings* and the fictional character Frodo) and with (explicit and implicit) parafictional information based on the content of the workspace (e.g. '$\Box_{LOTR}$trying-time(x)' and '(x)trying-time'). Importantly, not all propositional content of the workspace is updated as parafictional information simpliciter; Proper name conditions (e.g. 'Frodo(x)') are doubled and placed in the main box. This represents the fact that the abstract object Frodo also *exemplifies* being named 'Frodo' outside of *The Lord of the Rings*.[10]

If we adopt this strategy we add an abstract object to the shared ontology for any fictional entity that is introduced and is native to the relevant story. This means that we incorporate Zalta's metaphysical assumptions that entail the existence of abstract objects in the actual world. It also means that after reading *The*

---

[9] If we also incorporate theorem (I) we could simplify the representation of the common ground with respect to statements about native fictional characters.

[10] This move comes at a theoretical cost since it greatly complicates fictive closure (and opening), i.e. some discourse referents for ordinary objects in the workspace are replaced with discourse referents for abstract objects in the main box (and *vice versa* for fictive opening). The move is, however, in line with Zalta's analysis of fictional names in fictional statements (See Zalta [30,31]); Fictional statements do not involve reference to abstract objects but rather constitute the practice of story telling that determines – through an extended 'naming baptism' – what abstract objects the fictional names in parafictional and metafictional statements refer to. Moreover, fictive closure* and opening* allow an abstract object account to be extended to pronominal anaphora across mixed discourse with on the one hand a fictional statement and on the other hand either a metafictional or a parafictional statement.

Lord of the Rings the discourse referent for (the abstract object) 'Frodo' is accessible outside of the 'In *The Lord of the Rings*'-operator. To see how this solves the challenge posed by metafictional anaphora we first have to recognize that because Zalta draws a strong distinction between implicit and explicit parafictional statements, the challenge splits up in two sub-challenges: One of pronominal anaphora across mixed *explicit* parafictional/metafictional discourse and one of pronominal anaphora across mixed *implicit* parafictional/metafictional discourse. The central example up to this point, (5), is an example of pronominal anaphora across mixed *explicit* parafictional/metafictional discourse. I represent the workspace updated with (5) as follows:

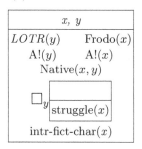

Next, we can rewrite (5) so that it is an example of pronominal anaphora across mixed *implicit* parafictional/metafictional discourse:

(10)   Frodo$_i$ goes through an immense mental struggle to save his$_i$ friends. Ah yes, he$_i$ is an intriguing fictional character!

I represent the workspace updated with (10) as follows:

$$
\begin{array}{|c|}
\hline
x,\, y \\
\hline
LOTR(y) \qquad \text{Frodo}(x) \\
A!(y) \qquad\quad A!(x) \\
\text{Native}(x,y) \\
(x)\text{struggle} \\
\text{intr-fict-char}(x) \\
\hline
\end{array}
$$

As the formalisms show, in both cases the discourse referent x for 'Frodo' is accessible outside of the 'In *The Lord of the Rings*'-operator. Hence the presupposition triggered by the pronoun 'he' in the metafictional statement in (5) and ((10) *can* be resolved.

Although this analysis seems to straightforwardly solve the problem of metafictional anaphora while avoiding the realist version of the problem of the wrong kind of object for implicit parafictional statements (i.e. abstract objects are the 'right kind of objects' to *encode* properties such as living in the Shire), the problem seems to reappear for explicit parafictional statements.[11] In Zalta's analysis, the explicit parafictional statement (2) expresses that the abstract object *The Lord of the Rings* encodes the vacuous property of being such that Frodo

---

[11] A similar concern has been voiced by Klauk [13].

exemplifies living in the Shire ('$\Box_{LOTR}$lives-in-Shire(f)'). That Frodo exemplifies living in the Shire ('lives-in-Shire(f)') is supposedly part of the content of (and true according to) *The Lord of the Rings*. However, the name 'Frodo' refers to an abstract object (i.e. Frodo is the abstract object that encodes all properties that Frodo exemplifies in *The Lord of the Rings* or '$f = ix(A!(x)\&\forall F((x)F \equiv \Box_{LOTR}F(f)))$'). Therefore it seems that we can derive that it is true in *The Lord of the Rings* that an abstract object exemplifies living in the Shire (i.e. we can derive '$\Box_{LOTR}$lives-in-Shire($ix(A!(x)\&\forall F((x)F \equiv \Box_{LOTR}F(f))))$'). This leads to a reoccurrence of the realist variant of the problem of the wrong kind of object; *The Lord of the Rings* is a story about flesh and blood hobbits, not about what properties abstract objects exemplify. Moreover, as Klauk [13] argues, although we may be able to imagine that an abstract object *exemplifies* living in the Shire, this amounts to imagining a category mistake (i.e. abstracts object cannot live in certain regions) which should be unusual and remarkable and cannot comprise our common practice of engaging in the content of a fictional work.

Actually, the above argument only works if we understand the '$\Box_{LOTR}$' as creating an extensional environment in which co-referring terms can be substituted *salva veritate*. Zalta (p.c.) suggests that the *LOTR* encoding prefix (and similar fiction prefixes) should be understood as creating an intensional environment similar to propositional attitude descriptions. Suppose John misread *The Lord of the Rings* and thinks Frodo actually exists. Statement (11):

(11)   John believes that Frodo lives in the Shire.

is a correct description of John's beliefs. Also, it is a fact that Frodo is the abstract object that encodes all properties that Frodo exemplifies in *The Lord of the Rings*. However, since 'John believes' creates an intensional environment this does not license us to infer that John believes that this abstract object lives in the Shire. Arguably, the same applies to (2); the *LOTR* encoding prefix creates an intensional environment so we can not derive from (2) and the fact that Frodo is the abstract object that encodes all properties that Frodo exemplifies in *The Lord of the Rings*, that *The Lord of the Rings* encodes that this abstract object exemplifies living in the Shire. Although this is a possible reply, this strategy seems to analyse 'being abstract' as a non-essential property (e.g. the abstract object Frodo lacks the property 'being abstract' in possible worlds compatible with John's beliefs) which seems counterintuitive.

Hence, although an abstract object account solves the sub-challenge of pronominal anaphora across mixed *implicit* parafictional/metafictional discourse, it may rerun into the realist variant of the problem of the wrong kind of object with the sub-challenge of pronominal anaphora across mixed *explicit* parafictional/metafictional discourse. Arguably, an analysis of the *LOTR* encoding prefix as creating in intensional environment solves this problem.

4.4   Dot-Object Theory: The Different Facets of Frodo

An alternative solution to the problem of metafictional anaphora that is formally similar to Zalta's solution is to follow Recanati [22] and claim that fictional names

in parafictional and metafictional statements refer to so-called 'dot-objects' that
are accessible in the main box.

Recanati goes back to Kripke's and Currie's intuition that fictional names
are ambiguous; In parafictional statements 'Frodo' refers to a flesh and blood
individual and in metafictional statements 'Frodo' refers to an abstract object.
Such an approach to (5) would result in the following DRS:

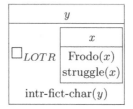

However, as explained in Sect. 1, this analysis seems to conflict with the
anaphoric link in (5). In other words, it is incompatible with what Recanati
dubs the 'Anaphora-Coreference Principle' (i.e. if a pronoun is anaphoric on an
antecedent name, the two terms co-refer) that is presupposed by previously discussed accounts of metafictional anaphora. However, Recanati argues, there are
apparent counterexamples to this principle. Take the following sentence:

(12)   Lunch$_i$ was delicious, but it$_i$ took forever. (adapted from Asher [1, p. 11])

The pronoun 'it' is anaphoric on the noun 'lunch' of the preceding clause. However, 'lunch' and 'it' do not co-refer; 'lunch' refers to food (which was delicious)
and 'it' refers to a social event (which took forever). Following Recanati, we
can save the Anaphora-Coreference Principle by appealing to the notion of a
dot-object (See e.g. Pustejovsky [21], Luo [16] or Asher [1]), i.e. "a complex
entity involving several 'facets'" [22, p. 15]. The noun 'lunch' is polysemous (i.e.
it can refer to food or a social event) and hence denotes a dot-object (represented as food • social event) involving several facets (i.e. a food facet and
a social event facet). Thus, in (12) 'lunch' and 'it' do actually co-refer (i.e to
the dot-object lunch), but the predicates 'being delicious' and 'taking forever'
apply to different facets of the object (i.e. respectively to the food facet and to
the social event facet). A common ground updated with (12) will thus look as
follows (where the dot-object lunch (x) is predicated over through its food facet
(x_1) and through its social event facet (x_2):

$$
\begin{array}{|c|}
\hline
x \\
\hline
\text{lunch}(x) \\
\text{DotObj}(x) \\
\text{delicious}(x_1) \\
\text{took-forever}(x_2) \\
\hline
\end{array}
$$

According to Recanati, fictional names are also polysemous (i.e. they can
refer to flesh and blood individuals or to abstract objects) and denote dot-objects
(e.g. 'Frodo' denotes the dot-object flesh and blood individual • abstract

object). In metafictional statements 'Frodo' refers to this dot-object through its abstract object facet. In (explicit and implicit) parafictional statements 'Frodo' refers to this dot-object through its flesh and blood individual facet.

Importantly, Recanati incorporates Zalta's distinction between encoding and exemplifying properties; The abstract object facet of Frodo both exemplifies properties such as being invented by Tolkien and encodes properties such as being a hobbit. Hence, the duality that is reflected in the two aspects of the dot-object Frodo, is also internal to the abstract object facet. Recanati agrees with Zalta that what properties the abstract object (facet) encodes is determined by our parafictional knowledge. However, whereas for Zalta the abstract object Frodo encodes just those properties that according to *The Lord of the Rings* the abstract object Frodo exemplifies; for Recanati, the abstract object facet of Frodo encodes just those properties that according to *The Lord of the Rings* the flesh and blood individual facet of Frodo has. Recanati thus maintains that the parafictional statement is *primarily* about the flesh and blood facet Frodo.

Applying Recanati's analysis to the workspace account suggests an adjustment of the fictive closure operation. I present a (simplified) representation of fictive closure** of (1):

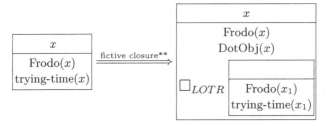

At fictive closure** we update the common ground with discourse referents for dot-objects for any newly introduced fictional character. These can be referred to as dot-objects (x), through their flesh and blood facet (x_1) (as is done in the parafictional condition) or through their abstract object facet (x_2). As in fictive closure* (See Sect. 4.3), proper name conditions are doubled and also placed in the main box.

A dot-object analysis of fictional characters solves the challenge posed by metafictional anaphora as in (5). A workspace updated with (5) looks as follows:

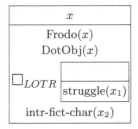

As the DRS shows the discourse referent x (predicated over through its abstract object facet in 'intr-fict-charx_2)') for 'Frodo' is accessible outside of the 'In *The*

Lord of the Rings'-operator.[12] Hence we can equate the discourse referents for 'Frodo' and 'he' and interpret the metafictional statement in (5). Although the solution is formally very similar to the solution offered by an abstract object account it avoids a reoccurrence of the realist variant of the problem of the wrong kind of object for explicit parafictional statements because in this analysis the name 'Frodo' refers to the dot-object Frodo *through its flesh and blood facet* rather than to an abstract object in both implicit and explicit statements.

Although a dot-object account of metafictional anaphora seems promising and avoids the difficulties of an abstract object account, some of the details still need to be worked out. First, as Recanati himself also notes [22, p. 43], the metaphysical status of dot-objects is controversial; Is a dot-object simply a pair of two facets, i.e. $x = \langle x_1, x_2 \rangle$? Must all facets of a dot-object exist in order for the dot-object to exist? These questions are especially pressing in the case of fiction where one of the facets of the dot-object (i.e. the flesh and blood individual facet) does not exist. In order to avoid metaphysical assumptions about the existence of multifaceted dot-objects, Recanati suggests that the correct objects of study are in fact dot-*concepts* (i.e. concepts of dot-objects) or mental files of dot-objects, rather than dot-objects. Similarly, in the DRS's above, the discourse referents and its associated conditions can be understood as representing concepts rather than objects (Cf. Maier [17]).

Second, it is not obvious how the crucial difference with an abstract object account (i.e. referring to a dot-object through different facets versus referring to an abstract object using different kinds of predicates), should be formalized. Whereas using the same argument (x) in the formalisation of metafictional and parafictional statements fails to show the difference, using different arguments (x_1 and x_2 as in the DRS's above) suggests that fictional names in parafictional and metafictional statements refer to distinct objects.

5   Conclusions

One of the aims of this paper has been to draw attention to the general challenge posed by anaphoric dependencies across different types of statements featuring fictional names. I have argued that the workspace account runs into difficulties with metafictional anaphora because the discourse referent for the fictional name introduced in the parafictional statement is not accessible outside of the 'In fiction F'-operator. This problem generalizes to other current dynamic approaches

---

[12] Here I assume that it is common ground that there is a dot-object Frodo after the parafictional update in (5). This is not obvious. Arguably, in the case of (12), as long as we talk about lunch as food and there is no mention of lunch as a social event, we are really just talking about lunch as food. Only at the introduction of the zeugmatic discourse does it become common ground that there is a dot-object (**food • social event**) that we refer to. The same could be true about (5). However, as Recanati suggests, the (overt or covert) 'In fiction F,'-prefix in parafictional discourse forces a 'metafictional perspective'; it makes us aware of the fictionality of the fictional characters and hence it is directly common ground at the parafictional update that we are referring to a dot-object including an abstract object facet.

that involve separation of the content and discourse referents of the fictional narrative. I have evaluated four different accounts of metafictional anaphora: A descriptivist approach (that requires an additional account of how to accommodate the right kind of definite descriptions), an approach based on changing the accessibility relations (that cannot straightforwardly be extended to a workspace account and runs into the antirealist variant of the problem of the wrong kind of object), an abstract object account (that potentially reruns into the realist variant of the wrong kind of object with explicit parafictional statements) and a dot-object account (that avoids the aforementioned problem but remains unclear on some crucial parts).

As mentioned in the introduction, any semantic account of fictional names will have to account for the use of pronominal anaphora across all acceptable types of mixed discourse. More specifically, it would be interesting to extend the described accounts of metafictional anaphora to certain problematic cases. For instance, suppose that apart from *The Lord of the Rings* Tolkien also wrote an alternative story (*The Lord of the Schmings*) in which the character Gimli (a dwarf in *The Lord of the Rings*) is an elf. I could then felicitously say (13):

(13)   In *The Lord of the Rings*, Gimli$_i$ is a dwarf but in *The Lord of the Schmings*, he$_i$ is an elf.

This is an example of pronominal anaphora across parafictional statements about different narratives. Arguably, although the pronoun 'he' is anaphoric on the name 'Gimli', the terms do not refer to the same fictional character Gimli since he is ascribed inconsistent (individual-level) predicates in the two different narratives. This is reminiscent of both the phenomenon of counterfictional imagination (See e.g. Friend [6]) and Geach's Hob-Nob puzzle:

(14)   Hob thinks a witch$_i$ blighted Bob's mare, and Nob thinks she$_i$ killed Cob's sow. (adapted from Geach [8])

Here the pronominal anaphora occur across two different propositional attitude reports and although the pronoun 'she' is anaphoric on 'a witch', there need not be one particular witch that is the object of thought of both Hob and Nob. Future research will have to determine how to account for (13) and determine its relation to other puzzles. Possible strategies in respectively an abstract object and a dot-object account would be to claim that 'he' and 'Gimli' in (13) refer to an abstract object that encodes inconsistent properties or that they refer to a dot-object with three or four different facets: two for the flesh and blood facets for Gimli the dwarf and Gimli the elf and one or two abstract object facets (depending on whether we allow for inconsistent abstract objects).

Acknowledgements. I thank Emar Maier, Edward Zalta and four anonymous reviewers for valuable comments that helped improve the paper. This research is supported by the Netherlands Organisation for Scientific Research (NWO), Vidi Grant 276-80-004 (Emar Maier).

Appendix

Examples of all six possible types of mixed discourse with fictional, parafictional and metafictional statements.

Type of mixed discourse	Example
Fictional/parafictional discourse	Robin Hood$_i$ stooped low and shot his$_i$ arrow just past the farmer's ear. Yes, that's right. In this story, he$_i$'s a villain!
Fictional/metafictional discourse	In order to capture the wicked witch, Mary$_i$ travelled to the woods and disguised herself as a potatoe. Yes, I know it's weird but she$_i$ was invented by a comedian.
Parafictional/fictional discourse	In the story that I'm about to tell you, a wizard named Brian$_i$ falls in love with a cauldron. This is how it goes: One day, he$_i$ was alone in his$_i$ study trying out a new love-potion recipe...
Parafictional/metafictional discourse	In *The Lord of the Rings*, Frodo$_i$ goes through an immense mental struggle to save his$_i$ friends. Ah yes, he$_i$ is an intriguing fictional character!
Metafictional/fictional discourse	I know this great story about a mythical creature named Frey$_i$. I will tell you. So one day she$_i$ was walking through the woods near her home...
Metafictional/parafictional discourse	Sherlock Holmes$_i$ is a fictional character created by Conan Doyle. In Conan Doyle's stories, he$_i$ is a private detective who investigates cases for a variety of clients, including Scotland Yard. (Adopted from Recanati [22, p. 37])

References

1. Asher, N.: Lexical Meaning in Context: A Web of Words. Cambridge University Press, Cambridge (2011)
2. Currie, G.: The Nature of Fiction. Cambridge University Press, Cambridge (1990)
3. Eckardt, R.: The Semantics of Free Indirect Discourse: How Texts Allow Us to Mind-Read and Eavesdrop. Brill Publishers, Leiden (2014)
4. Elbourne, P.: Situations and Individuals. MIT Press, Cambridge (2005)
5. Evans, G.: Pronouns, quantifiers and relative clauses (I). Can. J. Philos. **8**(3), 467–536 (1977)

6. Friend, S.: The great beetle debate: a study in imagining with names. Philos. Stud. **153**(2), 183–211 (2009)

7. Friend, S.: Fiction as genre. Proc. Aristot. Soc. **92**, 179–208 (2012)

8. Geach, P.: Intentional identity. J. Philos. **64**, 627–632 (1967)

9. Geurts, B.: Good news about the description theory of names. J. Semant. **14**, 319–348 (1997)

10. Heim, I.: E-type pronouns and donkey anaphora. Linguist. Philos. **13**, 137–177 (1990)

11. van Inwagen, P.: Creatures of fiction. Am. Philos. Q. **14**(4), 299–308 (1977)

12. Kamp, H.: A theory of truth and semantic representation. In: Groenendijk, J.A.G., Janssen, T.M.V., Stokhof, M.B.J. (eds.) Formal Methods in the Study of Language, Part 1, pp. 277–322. Blackwell Publishers Ltd., Oxford (1981)

13. Klauk, T.: Zalta on encoding fictional properties. J. Lit. Theor. **8**(2), 234–256 (2014)

14. Kripke, S.: Vacuous names and fictional entities. In: Philosophical Troubles: Collected Papers, vol. 1, pp. 52–74. Oxford University Press, Oxford (2011)

15. Lewis, D.: Truth in fiction. Am. Philos. Q. **15**(1), 37–46 (1978)

16. Luo, Z.: Formal semantics in modern type theories with coercive subtyping. Linguist. Philos. **36**(6), 491–513 (2012)

17. Maier, E.: Attitudes and mental files in Discourse Representation Theory. Rev. Philos. Psychol. **7**(2), 473–490 (2016)

18. Maier, E.: Fictional names in psychologistic semantics. Theor. Linguist. **43**(1–2), 1–45 (2017)

19. Mally, E.: Gegenstandstheoretische Grundlagen Der Logik Und Logistik. Barth, Leipzig (1912)

20. Matravers, D.: Fiction and Narrative. Oxford University Press, Oxford (2014)

21. Pustejovsky, J.: The Generative Lexicon. MIT Press, Cambridge (1995)

22. Recanati, F.: Fictional, metafictional, parafictional. Proc. Aristot. Soc. **118**(1), 25–54 (2018)

23. van der Sandt, R.A.: Presupposition projection as anaphora resolution. J. Semant. **9**(4), 333–377 (1992)

24. Semeijn, M.: A Stalnakerian analysis of metafictive statements. In: The Proceedings of the 21st Amsterdam Colloquium, pp. 415–424. (2017)

25. Stalnaker, R.C.: Pragmatics. Synthese **22**(1–2), 72–289 (1970)

26. Stokke, A.: Lying and asserting. J. Philos. **110**(1), 33–60 (2013)

27. Walton, K.L.: Mimesis as Make-Believe: On the Foundations of the Representational Arts. Harvard University Press, Harvard (1990)

28. Zalta, E.N.: Abstract Objects: An Introduction to Axiomatix Metaphysics. Springer, New York (1983). https://doi.org/10.1007/978-94-009-6980-3

29. Zalta, E.N.: Intensional Logic and the Metaphysics of Intentionality. MIT Press, Cambridge (1988)

30. Zalta, E.N.: Erzählung als taufe des helden: Wie man auf fiktionale objekte bezug nimmt. Zeitschrift fur Semiotik **9**(1–2), 85–95 (1987)

31. Zalta, E.N.: The road between pretense theory and abstract object theory. In: Everett, A., Hofweber, T. (eds.) Empty Names, Fiction and the Puzzles of Nonexistence, pp. 117–147. CSLI Publications, Stanford (2000)

Rule-Based Reasoners in Epistemic Logic

Anthia Solaki[(⊠)]

ILLC, University of Amsterdam, Amsterdam, Netherlands
a.solaki2@uva.nl

Abstract. In this paper, we offer a balanced response to the problem of logical omniscience, whereby agents are modeled as non-omniscient yet still logically competent reasoners. To achieve this, we account for the deductive steps that form the epistemic state of an agent. In particular, we introduce operators for applications of inference rules and design a possible-worlds model which is (a) equipped with a syntactic valuation, determining the agent's (explicit) knowledge, and (b) suitably structured by rule-induced transitions between worlds. As a result, we obtain a detailed analysis of the agent's reasoning processes. We then offer validities that exemplify how the problem of logical omniscience is avoided and compare our response to others in the literature. A sound and complete axiomatization is also provided. We finally show how simple extensions of this setting make it compatible with tools from Dynamic Epistemic Logic (DEL) and open to the incorporation of empirical findings on human reasoning.

Keywords: Rule-based reasoners · Epistemic logic ·
Dynamic epistemic logic · Logical omniscience · Bounded reasoning ·
Resource-bounded agents · Minimal rationality · Human reasoning

1 Introduction

Standard (S5) epistemic logic, using possible-worlds semantics, suffers from the *problem of logical omniscience* [13]: agents are modelled as reasoners with unlimited inferential power, always knowing whatever follows logically from what they know. This stark contrast with reality is also witnessed by experimental results indicating that subjects are systematically fallible in reasoning tasks [21,22]. It is even from a normative view that the standard account is insufficient, for it disregards the underlying reasoning of the agent and thus the restrictions on what can be *feasibly* asked of her. Therefore, knowledge should not be subject to logical closure principles. This, however, need not entail that agents are logically incompetent. While we often fail in complex inferences (e.g. due to lack of resources), we do engage in bounded reasoning: knowing that it is raining, and that we need a raincoat whenever it is raining, we do take a raincoat before leaving home. The empirical data also contributes to the case for logical competence, and as proposed in [9], we should seek a standard of *Minimal Rationality*. Drawing on these, we aim at modelling how an agent should *come to know* whatever can be feasibly reached from her epistemic state.

© Springer-Verlag GmbH Germany, part of Springer Nature 2019
J. Sikos and E. Pacuit (Eds.): ESSLLI 2018, LNCS 11667, pp. 144–156, 2019.
https://doi.org/10.1007/978-3-662-59620-3_9

In the twofold project of modelling a non-omniscient yet competent agent, we take on board the observations found in [7]. The deductive steps underpinning knowledge should be clearly reflected in an epistemic framework and this should still be compatible with "external" informational acts, as studied in DEL. We also place another desideratum: in principle, we should be able to employ empirical facts provided by cognitive scientists.

While many attempts have dealt with logical omniscience, not every attempt pursues a solution along the lines just described. Rule-based approaches, mainly applied on Artificial Intelligence, have paved the way towards our direction. Konolige [16] uses *belief sets* closed under an (incomplete) set of inference rules, but such (weaker) closure properties do not suffice to capture the agent's reasoning nor its cognitive load. Similar remarks apply to attempts which use modalities for reasoning processes [11], state-transitions due to inference [2–4], or arbitrary rule applications [14]. Collapsing reasoning processes to a modality, without a detailed analysis of their composition, would not help us determine what eventually makes them halt nor exploit investigations in psychology of reasoning which usually study *individual* inference rules on the grounds of cognitive difficulty. Interestingly, in [17], the author develops a logic where rules, accompanied by cognitive costs, are explicitly introduced in the language, but he gives no semantics, rendering the effect of his rule-operators unclear and the choice of axioms controversial.[1] Awareness settings [12] discern implicit and explicit attitudes, avoiding omniscience with respect to the latter, which additionally ask that agents are *aware* of a formula. Yet, an arbitrary syntactic awareness-filter cannot be associated with logical competence, and even if ad-hoc modifications are imposed (e.g. awareness closure under subformulas), forms of the problem are retained.[2]

The remainder is organized as follows: we first present our basic setting and explain how it contributes to the solution of the problem (Sect. 2). We then give a sound and complete axiomatization in Sect. 3 and in Sect. 4, we discuss how the basic framework can be easily adjusted to accommodate other directions and include sophisticated tools from logic and cognitive science.

2   The Setting

We first construct our logical language, building on the following definitions:

Definition 1 (Inference rule). *Given* $\phi_1, \ldots, \phi_n, \psi$ *in the standard propositional language* $\mathcal{L}_P$ *(based on a set of atoms* Φ*), an inference rule* R_i *is a formula of the form* $\{\phi_1, \ldots, \phi_n\} \rightsquigarrow \psi$.

---

[1] In [18] an impossible-worlds semantics is presented, but again reasoning is captured via modalities standing for a *number* of steps; this raises concerns analogous to the ones discussed before.

[2] A notable exception where awareness is affected by reasoning is given in [23]; in what follows, we design a rule-based approach but without appealing to a notion of awareness.

Notice that, according to this definition, each R_i stands for an *instance* of a rule (and not for a rule scheme). We then use $pr(R_i)$ and $con(R_i)$ to abbreviate the set of premises and the conclusion of R_i.[3] The rule is to say that whenever the premises are true, the conclusion is also true. We also use $\mathcal{L}_\mathcal{R}$ to denote the set of inference rules and $\mathcal{L} := \mathcal{L}_P \cup \mathcal{L}_\mathcal{R}$.

Definition 2 (Translation). *The translation of a formula in $\mathcal{L}$ is defined as:*
$Tr(\phi) := \phi$, *if $\phi \in \mathcal{L}_P$ and* $Tr(R_i) := \bigwedge\limits_{\phi \in pr(R_i)} \phi \to con(R_i)$, *if $R_i \in \mathcal{L}_\mathcal{R}$.*

We now define the language of this framework:

Definition 3 (Language $\mathcal{L}_{RB}$). *Given a countable set of propositional atoms Φ, the language $\mathcal{L}_{RB}$ is defined inductively as follows:*

$$\phi \quad ::= \quad p \quad | \quad \neg\phi \quad | \quad \phi \wedge \phi \quad | \quad K\psi \quad | \quad \langle R_i \rangle \phi$$

with $p \in \Phi, \psi \in \mathcal{L}, R_i \in \mathcal{L}_\mathcal{R}$.

As usual, $K\psi$ reads "the agent knows ψ". $\mathcal{L}_{RB}$ includes knowledge assertions for *rules* too. That is, apart of knowledge of facts, we can also express which *rules* the agent knows (and is therefore capable of applying). Each $\langle R_i \rangle$ is seen as a labeled operator for a rule-application. A formula $\langle R_i \rangle \phi$ reads "after some application of inference rule R_i, ϕ is true". Dual modalities of the form $[R_i]$ such that $[R_i]\phi$ expresses "after *any* application of R_i, ϕ is true", and the remaining Boolean connectives are defined as usual.

Next, we define our model motivated by the idea that reasoning steps, expressed through rule-applications, should be hardwired in it. We introduce possible worlds that are connected according to the effect of inference rules. Since an agent's reasoning affects the information she holds (rather than truth of facts), the usual valuation function is accompanied by a function yielding which formulas the agent knows at each world. In this sense, each world represents what is explicitly known at it and each rule triggers suitable transitions between them.

Definition 4 (Model). *A model is a tuple $M = \langle W, T, V_1, V_2 \rangle$ where*

- *W is a non-empty set of worlds.*
- *$T : \mathcal{L}_\mathcal{R} \to \mathcal{P}(W \times W)$ is a function such that a binary relation on W is assigned to each inference rule in $\mathcal{L}_\mathcal{R}$. That is, for $R_i \in \mathcal{L}_\mathcal{R}$, $T(R_i) = T_i \subseteq W \times W$, standing for the transition between worlds induced by the rule R_i.*
- *$V_1 : W \to \mathcal{P}(\Phi)$ is a valuation function assigning a set of propositional atoms to each world; intuitively those that are true at the world.*
- *$V_2 : W \to \mathcal{P}(\mathcal{L})$ is a function assigning a set of formulas of $\mathcal{L}$ to each world; intuitively those that the agent knows at the world.*

---

[3] We emphasize that R_i denotes a *single* rule instance. The rule, which is in fact a pair, composed of the set of premises and the conclusion, is given in terms of the notation $\rightsquigarrow$ for readability and convenience.

The truth clauses are given as follows:

Definition 5 (Truth clauses)

- $M, w \models p$ if and only if $p \in V_1(w)$ for $p \in \Phi$.
- $M, w \models K\phi$ if and only if $\phi \in V_2(w)$.
- $M, w \models \neg\phi$ if and only if $M, w \not\models \phi$.
- $M, w \models \phi \wedge \psi$ if and only if $M, w \models \phi$ and $M, w \models \psi$.
- $M, w \models \langle R_i \rangle \phi$ if and only if there exists some $u \in W$ such that $wT_i u$ and $M, u \models \phi$.

A formula is *valid in a model* if it is true at every world of the model and *valid* if it is valid in the class of all models. However, certain conditions have to be imposed on our initial, general class, to capture the desired effect of rule-applications. To that end, we need the following:

Definition 6 (Propositional truths). *Let M be a model and $w \in W$ a world of the model. Its set of propositional truths is $V_1^*(w) = \{\phi \in \mathcal{L}_P \mid M, w \models \phi\}$.*

We can now fix an appropriate class of models, denoted by **M**. For any model M (with $T(R_i) = T_i$ as defined above), $M \in \mathbf{M}$ if and only if:

1. For any inference rule $R_i = \{\phi_1, \ldots, \phi_n\} \rightsquigarrow \psi$, if $w \in W$ is such that $R_i \in V_2(w)$ and $\phi_1, \ldots, \phi_n \in V_2(w)$, then there exists a world $u \in W$ such that $wT_i u$.
2. For any $w, u \in W$ and inference rule $R_i = \{\phi_1, \ldots, \phi_n\} \rightsquigarrow \psi$, if $wT_i u$ then $R_i \in V_2(w)$, $\phi_1, \ldots, \phi_n \in V_2(w)$ and $V_2(u) = V_2(w) \cup \{\psi\}$.
3. For any $w \in W$ and $\phi \in \mathcal{L}$, if $\phi \in V_2(w)$ then $Tr(\phi) \in V_1^*(w)$.
4. For any $w, u \in W$ and inference rule R_i, if $wT_i u$ then $V_1^*(w) = V_1^*(u)$.

Condition 1 says that if a world represents an epistemic state containing the premises of a known rule R_i, then it must be connected to some other world by the corresponding T_i. Condition 2 says that if w is T_i-connected to u, then it must be that u enriches the epistemic state of w in terms of R_i. This is to ensure that each transition is associated with some addition of a conclusion to an epistemic state. Condition 3 is imposed to guarantee the veridicality of knowledge and the soundness of the known rules.[4] Finally, condition 4 states that T_i-connected worlds are propositionally indiscernible, i.e. transitions stand for purely epistemic actions.

We present some validities that illustrate desirable properties of reasoning processes and will be instrumental for a balanced response against logical omniscience. For notational convenience, we abbreviate sequences of rules as follows:

- $\langle \ddagger \rangle := \langle R_1 \rangle \ldots \langle R_n \rangle$
- $\langle \dagger \rangle := \langle R_1' \rangle \ldots \langle R_m' \rangle$

---

[4] Recall that $V_2 : W \rightarrow \mathcal{P}(\mathcal{L})$ and that $\mathcal{L} := \mathcal{L}_P \cup \mathcal{L}_R$. Moreover, it should be clear that the world u whose existence is guaranteed by condition 1, is such that it contains the conclusion of R_i, by condition 2, and the rule R_i is necessarily sound due to condition 3.

standing for "after some application of $R_1(R'_1)$, followed by some application of $R_2(R'_2)$, ..., followed by some application of $R_n(R'_m)$" (in that order). Similar abbreviations can be defined for the dual cases; for example, by using $[R_1], \ldots, [R_n]$ for the first sequence and $[R'_1], \ldots, [R'_m]$ for the second.

Theorem 1 (*M*-validities)

1. $\langle\ddagger\rangle K\phi \rightarrow Tr(\phi)$ *is valid in the class* **M**. *(Factivity)*
2. $\langle\ddagger\rangle K\phi \rightarrow \langle\ddagger\rangle[\dagger]K\phi$ *is valid in the class* **M**. *(Persistence)*
3. $\langle\ddagger\rangle K\phi \wedge \langle\dagger\rangle K\psi \rightarrow \langle\ddagger\rangle\langle\dagger\rangle(K\phi \wedge K\psi)$ *is valid in the class* **M**. *(Merge)*
4. *For any inference rule* R_i, $KR_i \wedge \bigwedge\limits_{\phi \in pr(R_i)} K\phi \rightarrow \langle R_i\rangle K con(R_i)$ *is valid in the class* **M**. *(Success)*

Proof

1. Take arbitrary model $M \in \mathbf{M}$ and arbitrary world $w \in W$ of the model. Suppose $M, w \models \langle\ddagger\rangle K\phi$. Unpacking the sequence according to the abbreviation, $M, w \models \langle R_1\rangle \ldots \langle R_n\rangle K\phi$, for the inference rules $R_1, \ldots, R_n$. Following Definition 5, there is a world $u_1 \in W$ such that wT_1u_1 and $M, u_1 \models \langle R_2\rangle \ldots \langle R_n\rangle K\phi$. Continuing like that, there is a world $u_n \in W$ such that $u_{n-1}T_nu_n$ and $M, u_n \models K\phi$, which in turn amounts to $\phi \in V_2(u_n)$. Then, by condition 3, $Tr(\phi) \in V_1^*(u_n)$. From condition 4, $Tr(\phi) \in V_1^*(u_{n-1})$. Continuing this process backwards, $Tr(\phi) \in V_1^*(w)$. Therefore $M, w \models Tr(\phi)$. Given the arbitrariness of $M \in \mathbf{M}$ and $w \in W$, we finally conclude that the formula is valid in the class **M**.

2. Take arbitrary model $M \in \mathbf{M}$ and arbitrary world $w \in W$ of the model. Suppose $M, w \models \langle\ddagger\rangle K\phi$. Unpacking the sequence according to the abbreviation, this amounts to $M, w \models \langle R_1\rangle \ldots \langle R_n\rangle K\phi$. As in the previous case, we obtain a chain $wT_1u_1 \ldots u_{n-1}T_nu_n$ such that $M, u_n \models K\phi$, which in turn amounts to $\phi \in V_2(u_n)$ (1). It suffices to show that $M, u_n \models [\dagger]K\phi$, i.e., by repeating the unpacking, now for $[\dagger] = [R'_1] \ldots [R'_m]$, that for every world $v_1 \in W$ such that $u_nT'_1v_1, \ldots$, for every world $v_m \in W$ such that $v_{m-1}T'_mv_m$, $M, v_m \models K\phi$, i.e. $\phi \in V_2(v_m)$. Take arbitrary such $v_1, \ldots, v_m$. Then due to condition 2 and (1), $\phi \in V_2(v_1)$ and continuing in the same fashion $\phi \in V_2(v_m)$. Therefore, $M, w \models \langle\ddagger\rangle[\dagger]K\phi$, hence $M, w \models \langle\ddagger\rangle K\phi \rightarrow \langle\ddagger\rangle[\dagger]K\phi$, as desired.

3. Take arbitrary model $M \in \mathbf{M}$ and arbitrary world $w \in W$ of the model. Suppose $M, w \models \langle\ddagger\rangle K\phi \wedge \langle\dagger\rangle K\psi$. So $M, w \models \langle\ddagger\rangle K\phi$ and $M, w \models \langle\dagger\rangle K\psi$. As above, we obtain a chain $wT_1u_1 \ldots u_{n-1}T_nu_n$ such that $M, u_n \models K\phi$, i.e. $\phi \in V_2(u_n)$, and a chain $wT'_1v_1 \ldots v_{m-1}T'_nv_m$ such that $M, v_m \models K\psi$, i.e. $\psi \in V_2(v_m)$. The rough idea of the proof is to make use of the conditions of **M** to merge the two chains. By condition 2, we know that $V_2(w) \subseteq V_2(u_n)$ and that $V_2(w)$ contains all the premises of rule R'_1, as well as the rule itself. Therefore, $V_2(u_n)$ in turn contains all the premises of rule R'_1 and the rule itself. By conditions 1 and 2, there is a world z_1 such that $u_nT'_1z_1$ and $V_2(z_1) = V_2(u_n) \cup \{con(R'_1)\}$. Now again, by condition 2, $V_2(v_1) = V_2(w) \cup \{con(R'_1)\}$ and since $V_2(w) \subseteq V_2(u_n)$: $V_2(v_1) \subseteq V_2(z_1)$, so we know that z_1 contains

the premises for R_2' and the rule itself. Again by conditions 1 and 2, there is a world z_2 such that $z_1 T_2' z_2$ and $V_2(z_2) = V_2(z_1) \cup \{con(R_2')\}$. Continuing like that, the alternations of condition 2 and condition 1, based on the initial assumptions, yield a world z_m such that $z_{m-1} T_m' z_m$ and $V_2(z_m) = V_2(z_{m-1}) \cup \{con(R_m')\}$ with $V_2(v_m) \subseteq V_2(z_m)$. Therefore $\psi \in V_2(z_m)$. In addition, as the constructed chain is of the form $u_n T_1' z_1 T_2' z_2 \ldots T_m' z_m$ and due to condition 2, $\phi \in V_2(z_m)$. So $M, z_m \models K\phi \wedge K\psi$, i.e. $M, u_n \models \langle \dagger \rangle (K\phi \wedge K\psi)$. So finally $M, w \models \langle \ddagger \rangle \langle \dagger \rangle (K\phi \wedge K\psi)$, as desired.

4. Take arbitrary model $M \in \mathbf{M}$ and arbitrary world $w \in W$ of the model. Suppose $M, w \models KR_i \wedge \bigwedge_{\phi \in pr(R_i)} K\phi$. Then $R_i \in V_2(w)$ and $\phi \in V_2(w)$, for every $\phi \in pr(R_i)$. Next, from conditions 1 and 2, there is $v \in W$ such that $wT_i v$ and $V_2(v) = V_2(w) \cup \{con(R_i)\}$. As a result, $M, v \models Kcon(R_i)$. Finally, $M, w \models \langle R_i \rangle Kcon(R_i)$, as desired.

Factivity says that whatever comes to be known is true, i.e. only true information or sound rules become known after reasoning, and *Persistence* says that it remains to be known throughout subsequent reasoning processes. *Merge* exemplifies how the agent merges different reasoning processes, thereby coming to know their outcomes. *Success* captures the effect of applying a rule: the conclusion is added in the agent's epistemic stack. As a concrete example, take the validity of

$$\bigwedge_{R_i = DNE, MP, CI} KR_i \wedge K\neg\neg\phi \wedge K(\phi \to \psi) \to \langle DNE \rangle \langle MP \rangle \langle CI \rangle K(\phi \wedge \psi):$$ after

successive applications of specific rules, namely *Double Negation Elimination* ($\{\neg\neg\phi\} \rightsquigarrow \phi$), *Modus Ponens* ($\{\phi, \phi \to \psi\} \rightsquigarrow \psi$) and *Conjunction Introduction* ($\{\phi, \psi\} \rightsquigarrow \phi \wedge \psi$), the agent's knowledge is gradually increased (Fig. 1).[5]

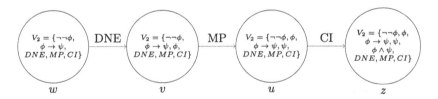

Fig. 1. A model where the reasoning steps of *Double Negation Elimination, Modus Ponens,* and *Conjunction Introduction,* taken in this order, correspond to transitions between worlds and reflect how the agent's knowledge is gradually increased.

Logical omniscience is indeed avoided in a balanced way, i.e. still escaping trivialized, totally ignorant agents. The values of knowledge assertions are determined by V_2, which need not obey any closure principle. On the other hand, suitable applications of inference rules, reflecting the effort to eventually reach a

---

[5] We use DNE, MP, and CI to label *particular* instances of Double Negation Elimination, Modus Ponens, and Conjunction Introduction – the ones indicated in parentheses. This labeling only serves the readability of the formulas.

conclusion, ensure that an agent can *come to know* consequences of her knowledge, provided that she follows the appropriate reasoning track. This is how we avoid an implausible commitment to an automatic and effortless way to expand one's knowledge, as the standard validity $K\phi_1 \wedge \ldots \wedge K\phi_n \rightarrow K\psi$ would dictate. Besides, Cherniak [9] emphasizes that we should view complex deductive reasoning as a task consisting of simple reasoning steps conjoined together. He also argues for a "well-ordering of inferences" in terms of their difficulty, depending both on the rule scheme in question and the logical complexity of its components. Similarly, according to Rips [20], deductive reasoning is a psychological procedure in which sets of formulas are connected via links, that essentially amount to applications of inference rules, just as our framework predicts. Overall, competence is preserved because we unfold the actual processes that result in knowledge and account for their dynamic nature. Logical ignorance is thus ruled out because of a more realistic modelling of the underlying reasoning and not because of ad-hoc restrictions imposed on an inflexible notion of knowledge.

It is interesting to see how our rule-based setting fits in the landscape of similar attempts. As in [1,14], temporal-style connections encode the progress in the agent's reasoning.[6] Unlike [4,11,14,18], we abstain from a generic notion of reasoning process, instead accounting explicitly for (a) specific rules available to the agent, (b) their individual applications, (c) their chronology, thus monitoring the path that eventually leads to knowledge. This elaborate analysis is, as we remarked above and will further discuss in Sect. 4, crucial in bridging epistemic frameworks with empirical facts.[7] Furthermore, the enterprise of providing a semantics contributes to Rasmussen's attempt [17], who keeps track of rules applied by the agent, on one hand, but lacks a principled way to assess the validity of his proposed axioms, on the other. Constructing a suitable semantic model that reflects rule-based reasoning gives a concrete view on the credibility of axioms and the adequacy of the solution. Finally, implicit and explicit notions can be discerned, not through an arbitrary filter (as with awareness), but through the analysis of the agent's reasoning.

3   Axiomatization

In this section, we develop the logic Λ_{RB}. We thus obtain a full-fledged logical response against the problem and solid ground to defend our selected axioms.

Definition 7 (Axiomatization of Λ_{RB}). *The axiomatization of Λ_{RB} is given by Table 1.*

Theorem 2 (Soundness). *The logic Λ_{RB} is sound with respect to $\mathbf{M}$.*

---

[6] We note that the frameworks described in [1–3] that extend the idea of state-transitions to multi-agent settings are particularly interesting for the development of multi-agent variants of our framework too.

[7] More on why this is a worthwhile task can be found in [6].

Table 1. Axiomatization of Λ_{RB}.

Axioms	
PC	All instances of classical propositional tautologies
K	$[R_i](\phi \to \psi) \to ([R_i]\phi \to [R_i]\psi)$
T	$K\phi \to Tr(\phi)$
Succession	$KR_i \wedge \bigwedge_{\phi \in pr(R_i)} K\phi \to \langle R_i \rangle \top$
Tracking knowledge	$\langle R_i \rangle K\chi \to \bigwedge_{\phi \in pr(R_i)} K\phi \wedge KR_i \wedge K\chi$, for $\chi \neq con(R_i)$
Knowledge of conclusions	$[R_i]Kcon(R_i)$
$Prop_1$	$\langle R_i \rangle \phi \to \phi$, for $\phi \in \mathcal{L}_P$
$Prop_2$	$\phi \to [R_i]\phi$, for $\phi \in \mathcal{L}_P$
Monotonicity	$K\chi \to [R_i]K\chi$
Rules	
Modus Ponens	From ϕ and $\phi \to \psi$, infer ψ
Necessitation	From ϕ infer $[R_i]\phi$

Proof. It suffices to show that the axioms of Definition 7 are valid in the class **M**, as our rules preserve validity as usual.

- The claim for PC and K is trivial.
- The claim for T follows immediately from condition 3.
- The claim for *Succession* follows from condition 1.
- For *Tracking knowledge*: Take any model $M \in \mathbf{M}$ and world $w \in W$ of the model such that $M, w \models \langle R_i \rangle K\chi$, for $R_i = \{\phi_1, \ldots, \phi_n\} \rightsquigarrow \psi$. So there is $u \in W$ such that $wT_i u$ and $\chi \in V_2(u)$. By condition 2, $\phi_1, \ldots, \phi_n, R_i \in V_2(w)$ and since $V_2(u) = V_2(w) \cup \{\psi\}$, $\chi \in V_2(w) \cup \{\psi\}$. So either $\chi \in V_2(w)$ or $\chi = \psi$. Finally, $M, w \models K\phi_1 \wedge \ldots \wedge K\phi_n \wedge KR_i \wedge K\chi$, for $\chi \neq \psi$.
- The claim for *Knowledge of conclusions* follows from condition 2.
- For $Prop_1$: Take any model $M \in \mathbf{M}$ and world $w \in W$ of the model such that $M, w \models \langle R_i \rangle \phi$ for $\phi \in \mathcal{L}_P$. Then, there is $u \in W$ such that $wT_i u$ and $M, u \models \phi$, i.e. $\phi \in V_1^*(u)$. By condition 4, $\phi \in V_1^*(w)$, i.e. $M, w \models \phi$ as desired.
- For $Prop_2$: Take any model $M \in \mathbf{M}$ and world $w \in W$ of the model such that $M, w \models \phi$. Take any $u \in W$ such that $wT_i u$. Then by condition 4, $\phi \in V_1^*(u)$, i.e. $M, u \models \phi$ so $M, w \models [R_i]\phi$, as desired.
- For *Monotonicity*: Take any model $M \in \mathbf{M}$ and world $w \in W$ of the model such that $M, w \models K\chi$, i.e. $\chi \in V_2(w)$. Take any $u \in W$ such that $wT_i u$. From condition 2, $\chi \in V_2(u)$, i.e. $M, u \models K\chi$. But then indeed $M, w \models [R_i]K\chi$.

Aiming at completeness, we follow the procedure of [8], employing *canonical models*.

Lemma 1 (Lindenbaum's Lemma). *If Γ is a Λ_{RB}-consistent set of formulas, then it can be extended to a maximal Λ_{RB}-consistent set Γ^+.*

Proof. The proof goes as usual in these cases. After enumerating $\phi_0, \phi_1, \ldots,$ the formulas of our language, one constructs the set Γ^+ as $\bigcup_{n \geq 0} \Gamma^n$ where: $\Gamma^0 = \Gamma$, $\Gamma^{n+1} = \Gamma^n \cup \{\phi_n\}$, if this is Λ_{RB}-consistent and $\Gamma^n \cup \{\neg\phi_n\}$ otherwise. The desired properties are easily obtained due to this construction.

Definition 8 (Canonical Model). *The canonical model $\mathcal{M}$ for Λ_{RB} is a tuple $\langle \mathcal{W}, \mathcal{T}, \mathcal{V}_1, \mathcal{V}_2 \rangle$ where:*

- $\mathcal{W} = \{w \mid w$ *a maximal Λ_{RB}-consistent set*$\}$.
- $\mathcal{T} : \mathcal{L}_{\mathcal{R}} \to \mathcal{P}(\mathcal{W} \times \mathcal{W})$, *such that for $R_i \in \mathcal{L}_{\mathcal{R}}$, $\mathcal{T}(R_i) = \mathcal{T}_i$, where $w\mathcal{T}_i u$ if and only if $\{\langle R_i\rangle\phi \mid \phi \in u\} \subseteq w$.*
- $\mathcal{V}_1 : \mathcal{W} \to \mathcal{P}(\Phi)$ *such that $\mathcal{V}_1(w) = \{p \in \Phi \mid p \in w\}$.*
- $\mathcal{V}_2 : \mathcal{W} \to \mathcal{P}(\mathcal{L})$ *such that $\mathcal{V}_2(w) = \{\phi \in \mathcal{L} \mid K\phi \in w\}$.*

It is easy to see that an equivalent formulation for the definition of $\mathcal{T}_i$ is $\{\phi \mid [R_i]\phi \in w\} \subseteq u$. Given the definition of the canonical model and our language $\mathcal{L}_{\mathrm{RB}}$, we show:

Lemma 2 (Existence lemma). *For any formula ϕ in our language and $w \in \mathcal{W}$, if $\langle R_i\rangle\phi \in w$ then there is $u \in \mathcal{W}$ such that $w\mathcal{T}_i u$ and $\phi \in u$.*

Proof. Suppose $\langle R_i\rangle\phi \in w$. Take $S = \{\phi\} \cup \{\psi \mid [R_i]\psi \in w\}$. This set is consistent. Were it inconsistent, there would be $\psi_1, \ldots, \psi_n$ such that $\vdash_{\Lambda_{\mathrm{RB}}} \psi_1 \wedge \ldots \wedge \psi_n \to \neg\phi$. Using $[R_i]$-necessitation, distribution and propositional tautologies we obtain $\vdash_{\Lambda_{\mathrm{RB}}} ([R_i]\psi_1 \wedge \ldots \wedge [R_i]\psi_n) \to [R_i]\neg\phi$. By the property of w as maximal consistent set and since $[R_i]\psi_1, \ldots, [R_i]\psi_n \in w$: $[R_i]\neg\phi \in w$. Therefore $\neg\langle R_i\rangle\phi \in w$. Indeed, we have reached a contradiction. Next, we extend S to S^+ according to Lindenbaum's lemma. Then, $\phi \in S^+$ and $[R_i]\psi \in w$ implies $\psi \in S^+$. Take $u := S^+$. As a result, $w\mathcal{T}_i u$ and $\phi \in u$.

Lemma 3 (Truth lemma). *For any formula ϕ in our language and $w \in \mathcal{W}$: $\mathcal{M}, w \models \phi$ if and only if $\phi \in w$.*

Proof. The proof is by induction on the complexity of ϕ.

- Base cases: Consider $\phi := p$ with $p \in \Phi$. Then $\mathcal{M}, w \models p$ if and only if $p \in \mathcal{V}_1(w)$, and by definition, this is the case if and only if $p \in w$. Next, take $\phi := K\psi$ with $\psi \in \mathcal{L}$. Then $\mathcal{M}, w \models K\psi$ if and only if $\psi \in \mathcal{V}_2(w)$, and by definition, this is the case if and only if $K\psi \in w$.
- For $\phi := \neg\psi$ and $\phi := \psi \wedge \chi$, the claim follows easily from I.H. and maximal consistency of w.
- For $\phi := \langle R_i\rangle\psi$ with I.H. that the result holds for ψ. Then $\mathcal{M}, w \models \langle R_i\rangle\psi$ if and only if there is $u \in \mathcal{W}$ such that $w\mathcal{T}_i u$ and $\mathcal{M}, u \models \psi$. By I.H. this is the case if and only if $\psi \in u$, and by definition of $\mathcal{T}_i$, we get $\langle R_i\rangle\psi \in w$. The other direction follows immediately from the existence lemma.

Theorem 3 (Completeness). *For any set of formulas Γ and formula ϕ in our language: $\Gamma \models_M \phi$ only if $\Gamma \vdash_{\Lambda_{\mathrm{RB}}} \phi$.*

Proof

- We first expand Γ to a maximal Λ_{RB}-consistent set Γ^+. Then, let the canonical model $\mathcal{M}$ be as constructed according to Definition 8. Then by Lemma 3, $\mathcal{M}, \Gamma^+ \models \Gamma$. It suffices to show that $\mathcal{M}$ fulfills the conditions of **M**.
- Condition 1 is satisfied.
  Take inference rule $R_i = \{\phi_1, \ldots, \phi_n\} \rightsquigarrow \psi$ and $w \in \mathcal{W}$ with $R_i, \phi_1, \ldots, \phi_n \in \mathcal{V}_2(w)$, i.e. $KR_i, K\phi_1, \ldots, K\phi_n \in w$ (1). We want to show that there is a world $u \in \mathcal{W}$ such that $w\mathcal{T}_i u$. From (1), $KR_i \wedge K\phi_1 \wedge \ldots \wedge K\phi_n \in w$. But from *Succession*, we get that $\langle R_i \rangle \top \in w$. Using the existence lemma, there is indeed $u \in \mathcal{W}$ such that $w\mathcal{T}_i u$.
- Condition 2 is satisfied.
  Suppose that $w\mathcal{T}_i u$ with $R_i = \{\phi_1, \ldots, \phi_n\} \rightsquigarrow \psi$, i.e. if $\phi \in u$ then $\langle R_i \rangle \phi \in w$. Take arbitrary $\chi \in \mathcal{V}_2(u)$. That is, $K\chi \in u$. Therefore, $\langle R_i \rangle K\chi \in w$. From *Tracking knowledge*, $\phi_1, \ldots, \phi_n, R_i \in \mathcal{V}_2(w)$. From *Knowledge of conclusions* and definition of $\mathcal{T}_i$, $K\psi \in u$, i.e. $\psi \in \mathcal{V}_2(u)$. Furthermore by this definition and *Monotonicity* we obtain that $\mathcal{V}_2(w) \subseteq \mathcal{V}_2(u)$. Therefore, $\mathcal{V}_2(w) \cup \{\psi\} \subseteq \mathcal{V}_2(u)$. Next take $\phi \in \mathcal{V}_2(u)$ with $\phi \neq \psi$. Then $\langle R_i \rangle K\phi \in w$. From *Tracking knowledge*, $K\phi \in w$. As a result, $\phi \in \mathcal{V}_2(w)$. Clearly then, $\mathcal{V}_2(u) = \mathcal{V}_2(w) \cup \{\psi\}$.
- Condition 3 is satisfied.
  Let ϕ be a formula in $\mathcal{L}$. Suppose that $\phi \in \mathcal{V}_2(w)$. That is, $K\phi \in w$. Then by T we obtain, $Tr(\phi) \in w$, that is $\mathcal{M}, w \models Tr(\phi)$ and therefore $Tr(\phi) \in \mathcal{V}_1^*(w)$.
- Condition 4 is satisfied.
  Take $w, u \in \mathcal{W}$ and $w\mathcal{T}_i u$. By definition of $\mathcal{T}_i$, if $\phi \in u$ then $\langle R_i \rangle \phi \in w$. Now take arbitrary $\phi \in \mathcal{L}_P$ such that $\mathcal{M}, u \models \phi$, i.e. $\phi \in \mathcal{V}_1^*(u)$. This means that $\phi \in u$, therefore $\langle R_i \rangle \phi \in w$. From *Prop*$_1$, we obtain $\phi \in w$, i.e. $\mathcal{M}, w \models \phi$ so $\phi \in \mathcal{V}_1^*(w)$. As ϕ was arbitrary, $\mathcal{V}_1^*(u) \subseteq \mathcal{V}_1^*(w)$. For the other inclusion, take arbitrary $\phi \in \mathcal{L}_P$ such that $\mathcal{M}, w \models \phi$, i.e. $\phi \in \mathcal{V}_1^*(w)$. This means that $\phi \in w$. From *Prop*$_2$, we get that $[R_i]\phi \in w$ too. Then we exploit the alternative definition of $\mathcal{T}_i$; since $[R_i]\phi \in w$, $\phi \in u$, i.e. $\mathcal{M}, u \models \phi$ so $\phi \in \mathcal{V}_1^*(u)$. As ϕ was arbitrary, $\mathcal{V}_1^*(w) \subseteq \mathcal{V}_1^*(u)$. Overall, $\mathcal{V}_1^*(w) = \mathcal{V}_1^*(u)$.

4   Extensions

This setting, whose key elements have been hitherto described, can also accommodate more intricate scenarios and facilitate applications informed by other disciplines. In particular, we briefly explain that other tools from (D)EL can be naturally combined with our rule-based logic and that, apart from AI, our syntactic approach can be also relevant for cognitive science.

First, a notion of *implicit* knowledge is not precluded in our framework, for it too employs possible worlds and can be easily endowed with an accessibility relation. Notions of belief can be also included along the lines presented so far, i.e. by simply attaching another function to the model, now yielding the explicit beliefs. Nevertheless, one might drop conditions on factivity or monotonicity. Regarding higher-order knowledge – provided that the language and the range of V_2 are extended – we can also avoid unlimited introspection, as is arguably desired for non-ideal agents. Just as with factual reasoning though, our framework can model *moderate* introspective abilities, via the introduction of introspective rules, whose semantic effect is similarly captured via world transitions.

Moreover, just like *public announcements* of DEL,[8] which may enhance the agent's knowledge, there can be actions for the learning of formulas in $\mathcal{L}$, that is not only of propositional formulas, but also of rules. Their effect is captured by (suitably) tweaking the components of our model to ensure that the formula or rule in question is included. In this way we can bring together external information and the agent's internal reasoning processes. For instance, consider an agent who knows $\phi \to \psi$ and $\neg\psi$, but comes to learn the *Modus Tollens* rule ($\{\phi \to \psi, \neg\psi\} \rightsquigarrow \neg\phi\}$). The combination of this learning action and the application of the rule leads to the agent coming to know that $\neg\phi$ too. Notice that the inclusion of DEL-style operators in this framework still allows for a sound and complete logic. This is because their effect is reducible to formulas not involving such operators. More specifically, *reduction axioms* can gradually "reduce" the truth of complicated formulas in the extended language to the truth of simpler formulas, up until the point where no operator is needed. Provided that these axioms are valid in **M**, we may simply refer to the completeness of Λ_{RB} and show that a logic built from Definition 7 and the reduction axioms is sound and complete w.r.t. **M**.

The use of labeled operators and the order-sensitivity of applications of rules make it easier to exploit the observations of cognitive scientists for a precise modelling of resource-bounded reasoners. For example, [15,20,22] suggest that not all rules are equally difficult for agents. According to Rips [20], the length and the difficulty of the rules involved in the mental proof constructed for a complex reasoning task determines its overall difficulty. In [19] the need to assign different weights to different rules is experimentally verified and in [24] empirically calculated weights are attached to different rules. Our framework can take these points into consideration. By fixing the agent's capacity (c), attaching empirically indicated weights to rules and introducing inequality formulas to the language (of the form $c \geq c_{R_i}$, where c_{R_i} is intuitively interpreted as the weight of R_i), we can place preconditions to applications of rules and therefore pinpoint where the cutoff of a reasoning process lies.[9]

On a more technical note, while we have presented a Hilbert-style axiomatization of Λ_{RB}, it would be interesting to develop a labeled sequent calculus alternative to this and investigate the proof-theoretic properties of our system. This

---

[8] As usual in DEL [5,10], we can add action operators to our language and capture their effect via model transformations triggered by the action. A formula with dynamic operators, of the form $[\alpha]\phi$, is evaluated by examining what the truth value of ϕ is at a transformed model, obtained via action α.

[9] In fact, this idea can be also pursued along the lines of DEL. The reasoning capacity c of the agent, as an additional component of our models, can be updated (i.e. reduced) following each rule application.

investigation can be especially relevant to the state-transition settings studying single- or multi-agent reasoning processes. In this way, we can obtain other independent technical results to motivate the use of such systems.

5   Conclusions

We argued that one of the important challenges for epistemic logic is not only to overcome logical omniscience, but to do so while securing the logical competence of agents. We located this endeavour's key parameter in bounded reasoning and spelled it out in logical terms by keeping track of the inference rules the agent applies. We explained how this enriches existing rule-based approaches and expands the scope of their applications. A sound and complete axiomatization was also provided, followed by a summary of our extensions of the core setting.

Acknowledgments. This work is funded by the Dutch Organization for Scientific Research, under the "PhDs in the Humanities" scheme (project number 322-20-018). The author also thanks the audience of the student session of ESSLLI 2018 and the anonymous reviewers for their valuable feedback.

References

1. Ågotnes, T., Alechina, N.: The dynamics of syntactic knowledge. J. Logic Comput. **17**(1), 83–116 (2007)
2. Ågotnes, T., Walicki, M.: A logic of reasoning, communication and cooperation with syntactic knowledge. In: AAMAS (2005)
3. Alechina, N., Jago, M., Logan, B.: Modal logics for communicating rule-based agents. In: ECAI (2006)
4. Alechina, N., Logan, B.: A logic of situated resource-bounded agents. J. Logic Lang. Inf. **18**, 79–95 (2009)
5. Baltag, A., Renne, B.: Dynamic epistemic logic. In: Zalta, E.N. (ed.) The Stanford Encyclopedia of Philosophy, Winter 2016 edn. Metaphysics Research Lab, Stanford University (2016)
6. van Benthem, J.: Logic and reasoning: do the facts matter? Stud. Logica: Int. J. Symbolic Logic **88**(1), 67–84 (2008)
7. van Benthem, J.: Tell it like it is: information flow in logic. J. Peking Univ. (Humanit. Soc. Sci. Edn.) **1**, 80–90 (2008)
8. Blackburn, P., de Rijke, M., Venema, Y.: Modal Logic. Cambridge University Press, New York (2001)
9. Cherniak, C.: Minimal Rationality. Bradford book, MIT Press, Cambridge (1986)
10. van Ditmarsch, H.P., van der Hoek, W., Kooi, B.P.: Dynamic epistemic logic and knowledge puzzles. In: Priss, U., Polovina, S., Hill, R. (eds.) ICCS-ConceptStruct 2007. LNCS (LNAI), vol. 4604, pp. 45–58. Springer, Heidelberg (2007). https://doi.org/10.1007/978-3-540-73681-3_4
11. Duc, H.N.: Reasoning about rational, but not logically omniscient, agents. J. Logic Comput. **7**(5), 633 (1997)
12. Fagin, R., Halpern, J.Y.: Belief, awareness, and limited reasoning. Artif. Intell. **34**(1), 39–76 (1987)

13. Fagin, R., Halpern, J.Y., Moses, Y.Y., Vardi, M.: Reasoning About Knowledge. MIT Press, Cambridge (1995)
14. Jago, M.: Epistemic logic for rule-based agents. J. Logic Lang. Inf. **18**(1), 131–158 (2009)
15. Johnson-Laird, P.N., Byrne, R.M., Schaeken, W.: Propositional reasoning by model. Psychol. Rev. **99**(3), 418–439 (1992)
16. Konolige, K.: A Deduction Model of Belief. Morgan Kaufmann Publishers, Burlington (1986)
17. Rasmussen, M.S.: Dynamic epistemic logic and logical omniscience. Logic Logical Philos. **24**, 377–399 (2015)
18. Rasmussen, M.S., Bjerring, J.C.: A dynamic solution to the problem of logical omniscience. J. Philos. Logic **48**(3), 501–521 (2019)
19. Rijmen, F., De Boeck, P.: Propositional reasoning: the differential contribution of "rules" to the difficulty of complex reasoning problems. Mem. Cogn. **29**(1), 165–175 (2001)
20. Rips, L.J.: The Psychology of Proof: Deductive Reasoning in Human Thinking. MIT Press, Cambridge (1994)
21. Stanovich, K.E., West, R.F.: Individual differences in reasoning: implications for the rationality debate? Behav. Brain Sci. **23**(5), 645–665 (2000)
22. Stenning, K., van Lambalgen, M.: Human Reasoning and Cognitive Science. MIT Press, Boston (2008)
23. Velázquez-Quesada, F.R.: Small steps in dynamics of information. Ph.D. thesis, Institute for Logic, Language and Computation (ILLC), Amsterdam, The Netherlands (2011)
24. Zhai, F., Szymanik, J., Titov, I.: Toward probabilistic natural logic for syllogistic reasoning (2015)

Free Relatives, Feature Recycling, and Reprojection in Minimalist Grammars

Richard Stockwell[✉]

University of California, Los Angeles, USA
rstockwell15@ucla.edu

Abstract. This paper considers how to derive free relatives—e.g. *John eats [DP what Mary eats]*—in Minimalist Grammars. Free relatives are string-identical to indirect questions—e.g. *John wonders [CP what Mary eats]*. An analysis of free relatives as nominalised indirect questions is easy to implement, but empirical evidence points instead to wh-words 'reprojecting' in free relatives. Implementing a reprojection analysis in Minimalist Grammars requires innovations to revise the stipulation that the probe always projects the head, and to allow features to be reused non-consecutively.

Keywords: Free relatives · Matching effects · Minimalist Grammars · Reprojection · Resource sensitivity

1 Introduction

This paper considers how to derive free relatives (FRs) (1) in Minimalist Grammars (MG) [24,26]. Section 2 illustrates MG with an analysis of indirect questions (IQs) (2), which are string-identical to FRs. An analysis of FRs as nominalised IQs is easy to implement, but the evidence presented in Sect. 3 points instead to wh-words 'reprojecting' in FRs [12]. In order to implement a reprojection analysis of FRs, I propose two innovations to MG in Sect. 4: one, a Reproject operation, revises the stipulation that the probe always projects the head; while a second, feature recycling, allows for features to be reused non-consecutively. I explore these innovations in Sects. 5 and 6 before concluding in Sect. 7.

(1) John eats [DP what he eats].

(2) John wonders [CP what Mary eats].

2 Minimalist Grammars, Indirect Questions, and Free Relatives

An MG analysis specifies a lexicon, pairing words with ordered lists of syntactic features. Matches between the first elements in these lists license applications of

© Springer-Verlag GmbH Germany, part of Springer Nature 2019
J. Sikos and E. Pacuit (Eds.): ESSLLI 2018, LNCS 11667, pp. 157–170, 2019.
https://doi.org/10.1007/978-3-662-59620-3_10

the structure building operations Merge and Move. We write $t[f]$ when the head of a tree—found by following the headedness arrows $>$ and $<$ down to a leaf node—has a sequence of syntactic features whose first element is f, and t for that tree with feature f erased. Merge (3) is licensed by matching category X and selector $=X$ features on the head of a pair of trees $t1$ and $t2$. If the selector $t1$ is lexical, it is linearized to the left $<$ and $t2$ is called the complement; otherwise $t1$ is linearized to the right $>$ and $t2$ is called the specifier. Move (4) is licensed by matching probe $+x$ and goal $-x$ features on a tree $t1$ containing a subtree $t2$. The probe $t1$ takes as a specifier the maximal projection of $t2$, $t2^M$, which is made phonetically null in its original position.[1] The matching features that license Merge and Move are 'checked' and deleted.

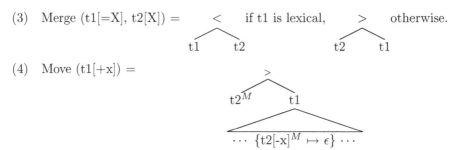

(3)   Merge $(t1[=X], t2[X]) = \qquad <$ if t1 is lexical, $>$ otherwise.

(4)   Move $(t1[+x]) =$

Two aspects of these definitions will be especially relevant for the analysis of FRs. First, projection: the selector in Merge and the probe in Move $t1$ projects the head, pointed to by $>$ and $<$, whose remaining features drive further structure building. Second, resource sensitivity: the matching features that license Merge and Move are checked and deleted, so are not available to drive any further structure building. While the second is a matter of technical implementation, the first is more central to Minimalist [8] reasoning in enforcing the endocentricity requirement of X-bar theory [6,18] and bare phrase structure [7]: each phrase has a head, one lexical item inside it that determines its distribution. However, MG Merge and Move are especially strict in stipulating the selector as the projector—projecting the selectee would still be endocentric.

To illustrate MG, an uncontroversial analysis of indirect questions (IQs) (2) can be implemented with the lexicon in (5), along with derivation (6) and derived (7) trees.

(5)   John :: D   Mary :: D   wonders :: =Q =D V   eats :: =D =D V
      ϵ :: =V C ϵ :: =V +wh Q   what :: D -wh

---

[1] Move is also subject to the shortest move constraint, which will not concern us here.

(6)

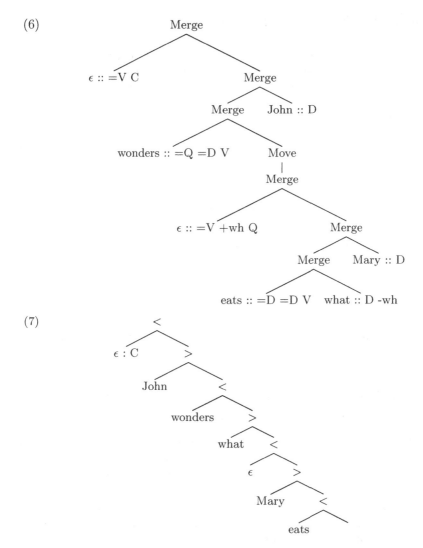

(7)

In two Merge steps, *eats* checks its $=D$ selector features against the category D features of *what*, then *Mary*. Omitting consideration of the tense layer, the phonologically null question complementiser $\epsilon ::= V + wh\, Q$ merges next. Move checks $+wh$ against $-wh$ to complete the indirect question, whose category Q is what *wonder* selects for. Construction of the main clause ends with the null complementiser $\epsilon ::= V\, C$ of the start category C.

The lexicon in (5) cannot derive free relatives (FRs) (1). Substituting *eats* for *wonder* does not converge, since *eats* selects for a complement of category D, not Q. A simple solution supplements the lexicon with $\epsilon ::= Q\, D$—a null D that selects a Q complement. Merge of $\epsilon ::= Q\, D$ with the output of Move in (6) converts the IQ from category Q to D, which *eats* can then take as its complement. Several versions of this null head analysis of FRs have been proposed, e.g. [15, 16]. However, as shown in the next section, there is a good deal of evidence against it.

3   Dual Role of Wh-Words in Free Relatives

MG Merge and Move conform to the general syntactic notion of headedness [6,7,18]. On the null head analysis, the derivation of a FR proceeds via an IQ of category Q. Projecting Q and merging $\epsilon ::= Q\, D$ seals off the wh-word inside the IQ, preventing it from informing the rest of the derivation; instead the null head will determine the FR's distribution. However, evidence suggests that the wh-word itself is the head of the FR, since the behaviour of a FR is keyed to the wh-word that forms it. This section shows that this is so for category distribution and case matching.

First, FRs distribute with the category of their wh-word (8), cf. [2]. A FR with *what* (1) distributes as a DP; but FRs formed with *where* distribute as PPs rather than DPs (8a), and those formed with *how* as AdvPs in not being able to intervene between a verb and its object (8b). On the null head analysis, this would require null, Q-complement-taking lexical items of many categories, e.g. $\epsilon ::= Q\, P$, $\epsilon ::= Q\, Adv$. And even then, nothing would enforce category matching between the null lexical item and the wh-word inside the FR, as required to rule out (9); in other words, we would expect mixtures like *what* $:: D$ *-wh* and $\epsilon ::= Q\, P$ to be grammatical.

(8)   a.   i.   Mary put the book [PP on the shelf] / [PP where she keeps it].

　　　　　　ii.   *Mary put the book [DP the shelf] / [DP what she built].

　　　b.   i.   John speaks [AdvP quickly] / [AdvP how you speak].

　　　　　　ii.   *John takes [AdvP quickly] / [AdvP how you write letters] notes.

(9)   *Mary put the book [PP ∅P [QP what_D John built]].

Second, in languages with morphological case, e.g. German [22], the wh-word in a FR must satisfy the case requirements of both the relative and matrix clauses. (10) is grammatical, since the nominative wh-word is the subject of the FR, which is the subject of the sentence. But (11) is ungrammatical due to the competing case requirements placed on the wh-word inside the FR, where it is an accusative object, and the FR as a whole, which is the nominative subject of the sentence. This conflict cannot be resolved—neither the accusative nor the nominative form of the wh-word will do. Since the null head analysis involves two distinct lexical items of category D—*what* and $\epsilon ::= Q\, D$—it offers no explanation for why they should match in case.

(10)   [DP_NOM Wer_NOM nicht stark   ist] muss klug   sein.
　　　　　　　　who   not   strong is   must clever be.

　　　'Who is not strong must be clever.'

(11)   *[DP_NOM { Wen_ACC } Gott schwach geschaffen hat] muss klug   sein.
　　　　　　　　　{ Wer_NOM }
　　　　　　　　　who   God weak   created   has must clever be.

　　　'Who God has created weak must be clever.'

Matches between FRs and their wh-words in distribution and case argue that the moving wh-word serves as the head of FRs, thereby determining their behaviour in the rest of the derivation. A number of researchers have reached this conclusion [2,10], a promising line of analysis being one where the wh-word itself projects the head of FRs [4,5,12].

The rest of this paper considers what it would take to modify the MG formalism to allow such a reprojection analysis to be expressed.[2] There are two issues to be overcome. First, while retaining endocentricity, a reprojection analysis directly contradicts the standard stipulation that a probe always projects over the goal it attracts to its specifier [1,9,11,21]—a stipulation that is baked into the definition of Move in (4). Second, even if the wh-word can be made the head, it will lack any features to drive the rest of the derivation. The next section proposes amendments to MG that allow a reprojection analysis of FRs to be implemented.

4   Implementing Reprojection in Minimalist Grammars

This section seeks to implement a reprojection analysis of FRs in MG. It does so by proposing two innovations: (i) Reproject, a new structure-building operation that revises the stipulation that the probe always projects; and (ii) feature recycling, a way for the category feature of the wh-word to be reused in the face of the resource sensitivity of MG. Consequences of these innovations and further directions are explored in Sects. 5 and 6.

4.1   Reproject

In revising the stipulation that a probe always projects, I propose to add Reproject to MG's inventory of structure building operations. We want Reproject to apply as in (12), reversing headedness to make *what* the head, thereby allowing *what* to determine the future of the derivation; and deleting the category feature Q, which would otherwise be left unchecked and cause the derivation to crash.[3]

(12)   Reproject

A general definition of Reproject is given in (13). A unary operation[4] applying to a tree with specifier *t1*, head *t2*, and complement *t3*, Reproject switches headedness to *t1* and deletes the category feature of *t2*, leaving *t3* unchanged.

---

[2] cf. Reprojecting head movement, e.g. [19], about which I will have nothing to say.
[3] The question of what features are on *what* in (12) is postponed to the next subsection.
[4] cf. [20] for a unary HPSG schema for free relatives in German.

$$(13) \quad \text{Reproject} \left(\begin{array}{c} \overset{>}{\diagup\diagdown} \\ t1 \quad \overset{<}{\diagdown} \\ \diagup\diagdown \\ t2[Y] \qquad t3 \end{array} \right) = \begin{array}{c} \overset{<}{\diagup\diagdown} \\ t1 \quad \overset{<}{\diagdown} \\ \diagup\diagdown \\ t2 \qquad t3 \end{array}$$

However, the way Y is checked without matching another feature in (13) would make Reproject very different from Merge and Move, which symmetrically check pairs of matching features.[5] Reproject is defined symmetrically in (14), where it applies to a tree where a reprojection feature $*Y$ on the specifier $t1$ matches the category of the head $t2$. Both features are checked, and headedness switches to $t1$. Using (14) means adding reprojection features $*Y$ to the inventory of syntactic features, and FR-specific reprojecting versions of wh-words to the lexicon; e.g. $what :: D$ -wh $*Q$, $where :: P$ -wh $*Q$.[6]

$$(14) \quad \text{Reproject} \left(\begin{array}{c} \overset{>}{\diagup\diagdown} \\ t1[*Y] \quad \overset{<}{\diagdown} \\ \diagup\diagdown \\ t2[Y] \qquad t3 \end{array} \right) = \begin{array}{c} \overset{<}{\diagup\diagdown} \\ t1 \quad \overset{<}{\diagdown} \\ \diagup\diagdown \\ t2 \qquad t3 \end{array}$$

However, as things stand the outcome of (14) has no features.[7] $t1$ is the head, but all its features have been checked en route to it becoming the specifier of $t2$. In deriving the FR in (1), $what :: D$ -wh $*Q$ has its D checked by Merge with $eats$, -wh by Move, and $*Q$ by Reproject, rendering its feature list empty, i.e. $what : \epsilon$. The next subsection proposes a way for category features to be reused so that the wh-word can serve as the head of FRs after Reproject.

4.2   Feature Recycling

After Reproject the wh-word is the head, ready to determine how the derivation will proceed. In order to account for the matching effects observed in Sect. 3, we would like $what$ to play a dual role in deriving (1) by contributing its category feature twice: first as complement to $eats$; then again after Move and Reproject to categorize the FR. However, MG structure building is resource sensitive: the matching features that license Merge and Move are checked and deleted. D of $what :: D$ -wh $*Q$ is expended in Merge with $eats$, and is subsequently unavailable.

---

[5] Even with persistent features [25], as discussed in the next subsection, while checking is not necessarily symmetric, structure building is still licensed by pairs of matching features.

[6] Wh-clustering [14]—see Sect. 5.3 below—provides a precedent for Reproject in being triggered by a feature on a specifier rather than a head. Clustering also involves complex specifiers, whereas I restrict attention here to trees with exactly one specifier.

[7] Recall from the illustration of MG in Sect. 2 that in order to converge, the derivation must reach the start category C.

Endowing *what* with a second *D* feature ordered after *-wh* and *$*Q$*, i.e.
what :: *D* *-wh* *$*Q$* *D*, (cf. *where* :: *P* *-wh* *$*Q$* *P*) invites the same empirical challenges as the null head analysis: with two separate category features, nothing
enforces category and case matching between FRs and their wh-words. Instead,
we would like one and the same *D* feature to contribute twice to the derivation.

Persistent features are an existing innovation that allow category features to
be used multiple times [25]. Merge continues to be licensed symmetrically by
matching features, but persistent features (underlined *$\underline{F}$*) do not have to delete.
Persistent features were motivated for implementing the movement theory of
control [17], allowing the same *$\underline{D}$* to occupy multiple argument positions by
satisfying multiple *=D* features. However, persistence in *what* :: *$\underline{D}$* *-wh* *$*Q$* does
not help in deriving FRs, since the two desired uses of *D* are non-consecutive.
Move and Reproject must apply after Merge of *what* with *eats* but before *what*
categorizes the FR. Hence *$\underline{D}$* would have to delete to allow Move to be triggered
by *-wh* before having the chance to provide the category of the FR.

I therefore propose feature recycling. Beyond persisting at the head of a list,
features can live on in the derivation by cycling to the end of the list after
licensing Merge. Feature recycling allows the *D* of *what* :: *D* *-wh* *$*Q$* to play a
dual but non-consecutive role, as shown in the derived tree of the FR from (1) in
(15). After licensing Merge with *eats*, *D* cycles to the end of the feature list; and
after Move checks *-wh* and Reproject checks *$*Q$*, the recycled *D* is back at the
head of *what*'s feature list to serve as the head of the FR. The diagram in (16)
summarizes the differences between standard resource sensitive feature checking,
persistent features, and feature recycling with respect to *what*'s *D* feature.

(15)

(16)

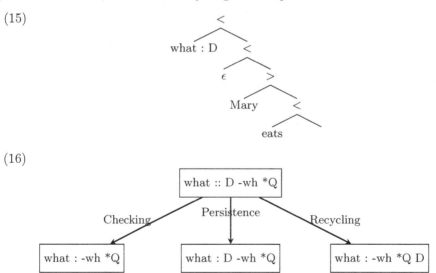

Thus we have implemented a reprojection analysis of FRs, cf. [4,5,12]. However, the analysis has come at the cost of two innovations. The first, Reproject, reverses headedness to the wh-word and deletes the category feature of the
embedded clause, which would otherwise be left unchecked and cause a crash.

The output of Reproject would lack any features were it not for the second innovation, feature recycling, which provides a way for the wh-word's category feature to be reused non-consecutively as the head of the FR. I explore these innovations further in the next two sections.

5   More on Reproject

This section considers the Reproject operation in greater detail, with discussion of the location of the triggering feature, Reproject's relationship to Move, and multiple-wh FRs.

5.1   Wh-Word Trigger

This subsection attempts to justify making the wh-word the trigger for Reproject. Whereas we could have put the triggering feature $*Y$ on the head that is reprojected over, the definition of Reproject in (14) has the triggering feature on the wh-word, e.g. *what* $:: D$ *-wh* $*Q$. The argument is based on the 'restrictor restriction' on FRs: wh-words can only form FRs if they lack a complement restrictor, as shown by the ungrammaticlity of (17).

(17)   *John eats [$_{DP}$ what food Mary eats].

The restrictor restriction is much easier to state if the wh-word is the trigger for Reproject. We can exclude from the lexicon wh-words with both a selector feature and a Reproject feature, e.g. *what* $:: =N\ D$ *-wh* $*Q$. By the time $\epsilon :: +\ wh\ Q$ interacts with the wh-word in the Move step, on the other hand, it would be unable to discriminate between a wh-word with or without a restrictor; in either case the wh-word will now have *-wh* as its first feature, $=N$ having long since been checked. Since the relevant information to distinguish between good FRs and (17) is not available to $\epsilon :: +\ wh\ Q$, the restrictor restriction on FRs argues that the wh-word should trigger Reproject.

However, this conclusion is provisional. Much stronger evidence would be cases of wh-words reprojecting over lexical items other than $\epsilon :: +\ wh\ Q$. Only then could we be sure that the wh-word is the trigger for Reproject, rather than the head reprojected over.

5.2   Reprojecting Move

Reproject as in (14) involves the heads *what* $:: D$ *-wh* $*Q$ and $\epsilon :: = V + wh\ Q$. It is licensed symmetrically by matching between the reprojection feature $*Q$ and category feature Q. But the two heads match in more than Q—they also match for *wh*. In Sect. 4.1, *wh* was checked by an application of Move before Reproject. But nothing said there enforces the co-occurrence in a lexical item's feature list of a Move licensee like *-wh* and a Reproject trigger like $*Q$. If Reproject is always fed by Move, we would be missing a generalization. Instead, we could recast Reproject as a version of Move, as in (18).

(18) Reprojecting Move (t1[+x Y]) =

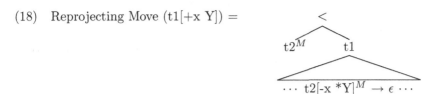

The definition in (18) enforces a dependency between Move licensees and Reproject triggers. Continuing to assume that the wh-word is the trigger for Reproject, we could further strengthen the dependency between moving and reprojecting by collapsing Move licensing and Reproject triggering into a single feature, *-x. The reprojecting lexical item *what* :: D *-wh would then trigger Reprojecting Move, as defined in (19).

(19) Reprojecting Move (t1[+x Y]) =

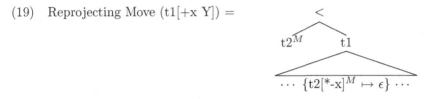

However, the definition in (19), reintroduces the asymmetry problem from our first definition of Reproject in (13). The category feature Y on *t1* is asymmetrically checked, deleting without having matched with another feature. Beyond this technical point, I cannot see how to decide between (18) and (19) as the definition of Reprojecting Move.

More generally, Reprojecting Move raises problems regardless of which of (18) or (19) we choose. For one, it increases the size of the 'moving window' on feature lists from one to two. Whereas Merge (3) and Move (4) apply based only on the first feature of the head, Reprojecting Move requires sight of the first two. A second problem might be that we have to restate all the restrictions on Move for Reprojecting Move, like the shortest move constraint and islands. Thus while reducing redundancy in enforcing a dependency between Move licensees and Reproject triggers, Reprojecting Move might increase redundancy elsewhere. However, redundancy only arises to the extent that movement as it feeds reprojection is subject to the same constraints as ordinary movement. If empirical investigation finds it to be subject to different constraints, we would have a strong argument for Reprojecting Move as its own operation. As with the previous subsection, we end by wondering whether wh-words reproject over lexical items beyond ϵ :: $+wh\,Q$.

5.3   Clustering and Reproject

Languages with overt multiple-wh movement to Spec, CP—like Bulgarian and Romanian (20) [3]—also allow multiple-wh FRs [23]:

(20) Ti-am   dat [ce   unde când a   trebuit instalat].
      CL2-have.1SG given   what where when has needed installed

      'I have given you the things that needed to be installed in the appropriate
      place at the appropriate time.'

Adopting the MG analysis of multiple wh-movement, or clustering, in [14],
only the topmost wh-word participates in a Move operation; the rest participate
in Cluster. Since multiple-wh FRs distribute with the category of the topmost
wh-word—D in (20)—the crucial dependency is between Reproject triggers and
Move licensees, not Cluster licensees.[8]

6   More on Reusing Features

Section 3 emphasised matches between FRs and their wh-words in order to argue
that the moving wh-word projects the head of FRs. Section 4.2 proposed feature
recycling as a way for the wh-word's category feature to be reused to implement
a reprojection analysis of FRs. This section explores ways in which a FR and
the wh-word that forms it can behave differently—if only very slightly. Slight
differences regarding case syncretism, complement/adjunct *where* FRs, and A-
bar features suggest that the notion we need may not be recycling but refreshing,
returning to the lexicon to pick another list of features compatible with the
morphological form of the word.

6.1   Case Syncretism

This subsection considers the case matching facts of Sect. 3 in greater detail. In
MG, abstract case licensing is implemented by k(ase) features, with all lexical
items of category D also bearing $-k$. The ungrammaticality of (21) shows that
$-k$ must recycle along with D in deriving FRs, since the FR as a whole must be
in a case position.

(21)   *It seems [DP what John eats] to be nice.

English wh-words do not differ morphologically for case,[9] which might suggest a generic $-k$ feature for English rather than more articulated $-nom$, $-acc$, etc.
Support for generic $-k$ comes from the lack of case matching effects in English
FRs. (22) is grammatical, despite the wh-word being assigned accusative internal to the FR, while the FR is nominative in the sentence overall. This contrasts
with the German mismatch from Sect. 3, repeated here as (23).

---

[8] Working with the definitions of Reproject in (14) and Cluster in [14], the lexical
   entries for the wh-words in (20) would be $ce :: D \bigtriangledown wh$ $-wh$ $*Q$, $unde :: P \bigtriangledown wh \bigtriangleup wh$
   and $când :: P \bigtriangleup wh$. The reprojection trigger $*Q$ would co-occur with $-wh$, not the
   Cluster licensor $\bigtriangledown wh$ or licensee $\bigtriangleup wh$.

[9] I set aside *whom* as an archaism.

(22) [$_{DP_{NOM}}$ What$_{ACC}$ John ate] killed him.

(23) *[$_{DP_{NOM}}$ { Wen$_{ACC}$ } Gott schwach geschaffen hat] muss klug   sein.
               Wer$_{NOM}$
              who   God weak   created   has must clever be.

'Who God has created weak must be clever.'

(24) [$_{DP_{NOM}}$ Wer$_{NOM}$ nicht stark   ist] muss klug   sein.
            who   not   strong is   must clever be.

'Who is not strong must be clever.'

For English, then, we can say that generic *-k* of *what* :: *D -k -wh *Q* is licensed inside the FR by a case-assigner, recycled along with *D*, and licensed again by another case-assigner in the main clause. In German, on the other hand, the morphological differences among wh-words for case might suggest lexical items of category *D* differ among *-nom*, *-acc*, etc. In (24), *wer* :: *D -nom -wh *Q* is licensed for nominative inside the FR, recycled with *D* in forming the FR, and licensed again for nominative in the main clause.

Switching between *-acc* and *-nom* is ungrammatical in (23); but this is not always the case in German for a FR configuration that mixes nominative and accusative. Such mismatches are possible when there is case syncretism, as in (25) [22], where neuter gender *was* can realise either nominative or accusative.

(25) [$_{DP_{NOM}}$ Was$_{ACC}$ du   gekocht hast] ist schimmlig.
          What   you cooked have   is moldy.

'What you have cooked is moldy.'

Starting with *was* :: *D -acc -wh *Q* inside the FR, we cannot switch to *was* :: *D -nom -wh *Q* in the main clause via feature recycling, incorrectly predicting (25) to be bad. We could salvage feature recycling by changing our assumptions about case features, claiming that there is only one lexical item *was* :: *D -nomacc -wh *Q* whose underspecified *-nomacc* case feature can be checked by either a nominative or accusative case-assigner.

Alternatively, we could account for (25) by refreshing rather than recycling the features of *was*. After moving to specifier position and reprojecting, *was* :: *D -acc -wh *Q* has exhasted its list of features. Rather than pre-empting this problem with feature recycling, *was* : ϵ could refresh its features by reaching back into the lexicon for a list of features compatible with its morphological form. This reuse of the string rather than its features allows for the refreshed feature list to be slightly different—including *-nom* rather than *-acc* in deriving (25).[10]

While either underspecified *-nomacc* or feature refreshing would account equally well for German case syncretism, refreshing appears to be the only plausible option for the topic of the next subsection.

---

[10] The length of the reused string is finitely bounded, in that only lexical items and not phrases can be reused – recall the restrictor restriction on FRs from Sect 5.1.

6.2   Complement vs. Adjunct *Where* Free Relatives

The previous subsection showed that syncretism allows wh-words and the FRs they form to differ in case. This section argues that sycretism also allows differences in category.

Following a prominent analysis of adjunction in MG [13], PPs have very different categories depending on whether they appear in complement or adjunct position: $where :: P$ -*wh* is a complement to verbs like $put :: =P =D =D\ V$, whereas $where :: \approx V$ -*wh* adjoins to category V, which continues to be the head. In (26), *where* is an adjunct to *eats* inside the FR, while the FR as a whole is a complement to *put*. Thus *where* has different category features internal and external to the FR, which would not follow from feature recycling.

(26)   Mary put the book [$_{\text{PP}_{\text{COMP}}}$ where$_{\text{PP}_{\text{ADJ}}}$ John eats].

It is difficult to countenance an underspecification analysis among two different feature lists for *where* along the lines of underspecified -*nomacc* in the previous subsection. That leaves us with feature refreshing: in deriving (26), $where :: \approx V$ -*wh* is exhausted to $where : \epsilon$ in deriving the FR, before refreshing as $where :: P$ -*wh* for the main clause.

6.3   A-bar Features

However feature reuse is to be characterised – whether features are recycled or refreshed – A-bar features are not reused. Despite being headed by a wh-word, a FR cannot itself undergo wh-movement, as in (27).

(27)   *[$_{\text{DP}}$ What John eats] does Mary eat t?

Unlike category and case features, which play a dual role in deriving FRs, -*wh* is definitively consumed in moving *what* inside the FR, so would have to be barred from recycling. In terms of refreshing, meanwhile, we could say that features are refreshed based on the non-wh part of the word, assuming decomposition of e.g. German *wer* into wh *w-* + -*er* nominative D.

Yet FRs can embark on other A-bar movements, e.g. topicalisation in (28).

(28)   [$_{\text{DP}}$ What John eats], I eat t.

Still, the movement in (28) cannot result from reusing a feature. Assuming topicalisation is licensed by -*top*, it must be added to the FR after it has been fully formed: while the FR as a whole is topicalised in (28), the wh-word does not undergo topicalisation inside the FR, so -*top* cannot have been present on *what* at the start of the derivation. The opposite behaviour of -*wh* and -*top* in being active only internal vs. external to the FR tracks the difference between intrinsic vs. optional features [8, p. 231].

7 Conclusion

This paper set out to derive FRs in MG. Reviewing category and case matching effects motivated implementing a reprojection analysis. Doing so came at the cost of two innovations. Reproject, a new structure-building operation, revised the stipulation that the probe always projects. Answers to the technical questions of whether the trigger is the wh-word, and whether Reproject is a special case of Move, depend on the empirical question of whether wh-words reproject over lexical items other than $\epsilon :: + wh\,Q$. The second innovation—feature recycling—provided a way for the category feature of the wh-word to be reused nonconsecutively as the head of the FR in the face of the resource sensitivity of MG. The slight relaxation of matching effects where there is syncretism suggested features might be refreshed rather than recycled, though A-bar features cannot be reused.

References

1. Boeckx, C.: Bare Syntax. Oxford University Press, Oxford (2008)
2. Bresnan, J., Grimshaw, J.: The syntax of free relatives in English. Linguist. Inq. **9**(3), 331–91 (1978)
3. Caponigro, I., Fălăuş, A.: The functional nature of multiple wh-free relative clauses in Romanian. Poster Presented at SALT 28, MIT (2018)
4. Cecchetto, C., Donati, C.: Relabeling heads: a unified account for relativization structures. Linguist. Inq. **42**, 519–560 (2011)
5. Cecchetto, C., Donati, C.: (Re)labeling. MIT Press, Cambridge (2015)
6. Chomsky, N.: Remarks on nominalization. In: Jacobs, RA., Rosenbaum, D.H. (eds.) Reading in English Transformational Grammar, pp. 184–221. Ginn, Waltham (1970)
7. Chomsky, N.: Bare phrase structure. In: Webelhuth, G. (ed.) Government and Binding Theory and the Minimalist Program, pp. 385–439. Blackwell, Oxford (1995)
8. Chomsky, N.: The Minimalist Program. MIT Press, Cambridge (1995)
9. Chomsky, N:. On phases. In: Foundational Issues in Linguistic Theory: Essays in Honor of Jean-Roger Vergnaud, pp. 133–166 (2008)
10. Citko, B.: Missing labels. Lingua **118**(7), 907–944 (2008)
11. Collins, C.: Eliminating labels. In: Epstein, S., Seely, T. (eds.) Derivation and Explanation in the Minimalist Program, pp. 45–61. Blackwell, Oxford (2002)
12. Donati, C.: On wh-head movement. In: Wh-Movement: Moving on, pp. 21–46 (2006)
13. Frey, W., Gärtner, H.M.: On the treatment of scrambling and adjunction in minimalist grammars. In: 2002 Proceedings Formal Grammar (2002)
14. Gärtner, H.M., Michaelis, J.: On the treatment of multiple-wh-interrogatives in minimalist grammars. In: Language and Logos, pp. 339–366 (2010)
15. Groos, A., Van Riemsdijk, H.: Matching effects in free relatives: a parameter of the core grammar. In: Theory of Markedness in Generative Grammar, pp. 171–216 (1981)
16. Grosu, A.: Three Studies in Locality and Case. Routledge, London (1994)
17. Hornstein, N.: Movement and control. Linguist. Inq. **30**(1), 69–96 (1999)

18. Jackendoff, R.: X-bar-Syntax: A Study of Phrase Structure. MIT Press, Cambridge (1977)
19. Koeneman.: The flexible nature of verb movement. Ph.D. dissertation, Universiteit Utrecht (2000)
20. Müller, S.: An HPSG-analysis for free relative clauses in German. Grammars **2**(1), 53 (1999)
21. Pesetsky, D., Torrego, E.: Probes, goals and syntactic categories. In: Otsu, Y. (ed.) Proceedings of the seventh annual Tokyo conference on psycholinguistics, pp. 25–60. Hituzi Syobo, Tokyo (2006)
22. van Riemsdijk, H.: Free Relatives. The Blackwell Companion to Syntax. Blackwell, Oxford (2007)
23. Rudin, C.: Multiple wh-relatives in Slavic. In: Compton, R., Goledzinowska, M., Savchenko, U. (eds.) FASL, pp. 282–307 (2007)
24. Stabler, E.: Derivational minimalism. In: Retoré, C. (ed.) LACL 1996. LNCS, vol. 1328, pp. 68–95. Springer, Heidelberg (1997). https://doi.org/10.1007/BFb0052152
25. Stabler, E.: Sidewards without copying. In: Formal Grammar, vol. 11, pp. 133–146 (2006)
26. Stabler, E.: Computational perspectives on minimalism. In: Boeckx, C. (ed.) Oxford Handbook of Linguistic Minimalism, pp. 617–642 (2011)

Playing with Information Source

Velislava Todorova[(✉)]

Sofia University, Sofia, Bulgaria
todorova.slava@gmail.com

Abstract. In this paper I present a NetLogo simulation program which models human communication with indication of information source. The framework used is evolutionary game theory. Under different initial settings the individuals in the simulation either learn to systematically indicate their information source or not. I use several examples to show how this difference is connected to the impact of one's speech behaviour on their reputation. In a community where this impact is high, the individuals who do not mark their information source lose reputation quickly and are ultimately excluded from the community. My hope is that this simulation program can help understand better the grammatical category evidentiality – the prototypical way of systematically indicating information source – and also shed some light on the question why this category developed in some languages and not in others.

Keywords: Information source · Simulation ·
Evolutionary game theory · Evidentiality

1 Introduction

Every language has a way of indicating the information source. If this way is a special grammatical category, it is called *evidentiality*. If it is a special use of a category with a different primary meaning, it would be rather called an *evidential strategy*. And if the marking is done by lexical means, it would be simply a *lexical expression of information source*.[1] There are even further means to indicate one's source: for example, the scientific community has developed efficient and highly conventionalized, yet not properly linguistic, ways to make bibliographical references.

I have created a simulation program[2] that models human communication with a focus on information source indication. The simulation is not meant to represent specifically the linguistic marking of information source, but its main motivation is to shed light on the possible reasons for the appearance of evidentiality in some languages and not in others.

---

[1] For a clear distinction between the possible ways to indicate information source, see (Aikhenvald 2004, esp. Sect. 1.2.2).

[2] It could be viewed and downloaded from https://github.com/SlavaTodorova/InformationSourceSimulation.git.

© Springer-Verlag GmbH Germany, part of Springer Nature 2019
J. Sikos and E. Pacuit (Eds.): ESSLLI 2018, LNCS 11667, pp. 171–184, 2019.
https://doi.org/10.1007/978-3-662-59620-3_11

The intuition behind the simulation scenario is that the indication of information source is connected to the reputation of speakers. In Aikhenvald's (2004, p. 359) words:

> In a small community everyone keeps an eye on everyone else, and the more precise one is in indicating how information was acquired, the less the danger of gossip, accusation, and so on. No wonder that most languages with highly complex evidential systems are spoken by small communities.

This article will show how reputation, and most precisely the impact of one's speech on their reputation, does indeed play a role in the development of a systematic practice of marking information source.

2   Structure of the Simulation

Before the start of the simulation, the user specifies the number of individuals in the population, the number of witnesses, the level of reliability of the information and the impact of the speaker's messages on their reputation. When the simulation starts, an event takes place and some individuals witness it. The witnesses might get a wrong impression of the event,[3] but either way they search for hearers to share what they think has happened. If there are uninformed individuals near the witness and if those individuals find the reputation of the witness high enough, a conversation begins. In the conversation the speaker utters a message reporting the belief they have and, optionally, marking the information source. Hearers either believe what they have heard or not, and decide if the information should be spread further. There is again the chance of misunderstanding the message.

When the whole population has been informed (or misinformed) about the event, all individuals observe, as by providence, whether their believes and statements are true or false. On the basis of these observations their strategies (to prefer one message or another, and to rather believe or disbelieve a message) are adjusted, and their reputation levels are changed. With this a step in the simulation is completed, and a new one can start, with a new event and new witnesses.

At the end of each step of the simulation, the individuals with minimal (zero) reputation are excluded from the community. If the remaining individuals are less than the number specified by the user, and if there are individuals with maximal reputation, then a new member is added to the community. This new member has exactly the same strategy as one (a random one) of the individuals with maximal reputation.

---

[3] For the sake of simplicity, in this simulation all agents are assumed to be cooperative and benevolent. This means that there would be no liars in the community. Still, in order to bring the model closer to reality there will be a chance of misunderstanding, which will result in formulation and spread of false information.

3 The Game

3.1 Players and Moves

The simulation is a game in the sense of evolutionary game theory[4] and Fig. 1 presents its extensive form. At the beginning, Nature (Player 0)[5] gives firsthand evidence to some of the players. Firsthand evidence can be interpreted correctly or incorrectly. As it is not a conscious decision the player makes, I assume it is again Nature's choice. The player cannot be sure if the belief they formed is true or false.[6] They nevertheless have a belief and search for a hearer to share it. If a hearer is found, they would be Player 2, and Player 1, the speaker, would choose either the basic message to communicate the information, or a more complicated message, marked for information source, viz. a firsthand information message.[7] I assume that the speaker chooses a message that correctly represents their belief and the only difference in the possible messages is that one is marked for information source and the other is not. Then Player 2 decides whether to believe the information. In the end both players have some utility from the conversation: in the leaves of the tree the first number is always the speaker's utility and the second one is the hearer's.

The second branch of the game tree – the hearsay subgame – starts with Nature giving hearsay evidence to a player.[8] The player might be given a true or false piece of information, but they cannot distinguish between the two cases. They have decided according to their hearer strategy (when they were Player 2 in the firsthand evidence subgame) if they will believe or doubt the information.[9] If they believe it, it can turn out that they have misunderstood.

There are two options for the case in which Player 1 has formed a belief – they can either use the basic message or a message marked for hearsay information.[10]

---

[4] I am taking here a broad sence of the term'evolutionary game theory', one which includes learning game theory. I do this following (Mühlenbernd 2011), as my work is closely based on his approach.

[5] Nature is a fictitious player in the game, whose actions are those choices that do not depend on either of the two actual players.

[6] The information sets (the sets of states between which a player cannot distinguish) are represented in the tree by dotted arcs.

[7] To give an example, in English the difference between these two kinds of messages would be the distinction between "It is raining" and "I see that it is raining."

[8] The hearsay information is given to players by other players in a previous stage of the same game. However, the structure of the simulation is such that whether a player will get hearsay information, is decided together with the distribution of firsthand evidence – all the individuals who didn't receive firsthand evidence, have to eventually be informed by others.

[9] Technically, the application of the hearer strategy takes place in the previous stage of the game, when Player 1 in this second branch has been Player 2 in the first branch. However I repeat this part of the game, as it is important to distinguish between the states that result from different outcomes in the previous stage.

[10] An example from English for the difference between a message of the basic type and a message of the hearsay type would be the same as between the sentences "It is raining" and "They say it is raining."

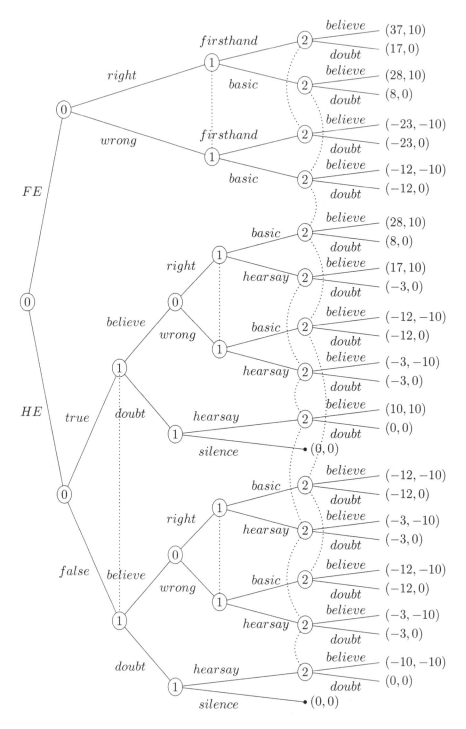

Fig. 1. Extensive form of the game. (Outcomes are calculated for reputation cost/gain values of 20 and 10 for the firsthand and the basic message respectively.)

The firsthand information message cannot be used, as its sincerity condition requires the additional belief that the speaker has witnessed the event. There is no possibility for insincerity in the model (lies are not allowed), so firsthand information messages are excluded in the hearsay scenario and similarly the hearsay information messages are excluded in the firsthand evidence scenario.

In the case in which Player 1 has not formed a belief, the options are to either pass along the information using the hearsay information message, or to stay silent. The hearsay information message is the only admissible one here, as all other would not be sincere given that the speaker does not believe the information is true.[11]

Here again, just like in the firsthand evidence subgame, Player 2 has to choose if to believe the message they hear. They cannot tell if a speaker uttering the basic message was a witness or not, nor if the information this message carries is true.

There is one move by Nature that is omitted in the tree for simplicity. After Player 2 decides to believe Player 1, it could turn out that they have misunderstood and formed a false belief. In this case neither the speaker nor the hearer gains or loses anything and their strategies are not updated, since neither the speaker may draw a conclusion about the persuasive power of their message, nor the hearer can blame the negative outcome of the communication on their naivity.

3.2   Outcomes, Costs and Gains

After each conversation, both the speaker and the hearer receive some utility. The precise value of the received utility depends on the perlocutionary goal the speaker had, the complexity of the message employed, the reaction of the hearer and, ultimately, on the truth of the information transmitted.

The basic outcome of the communication – the one dependent of the truth of the information – is positive for both players, if true information has been shared and believed, and negative if false information has been shared and believed. In the cases when a piece of information (true or false) is not believed, there is a neutral outcome. Table 1 presents the basic outcomes.

The basic outcome is the only factor to be considered for the hearer's utility. For the speaker there are other relevant factors. One of them is the perlocutionary goal.

In line with Martina Faller's discussion of the purposes of conversations with different evidentials in Quzco Quetchua (Faller 2006, pp. 28–29), I assume that whenever the speaker does have a belief, their goal is to persuade the hearer;

---

[11] It is clear that in English a sentence of the form "They say it is raining" can be sincere even if the speaker is convinced it is not raining. Languages that do not use such embedding structures, but grammatical evidentiality also seem to allow for the sincere utterance of hearsay marked messages even when the speaker knows the information is false. For an example from Bulgarian, see (Smirnova 2011, p. 27) and for one from Quechua, see (Faller 2006, p. 4).

Table 1. Basic outcomes

	Player 1 (Speaker)	Player 2 (Hearer)
Believed true information	10	10
Not believed true information	0	0
Believed false information	−10	−10
Not believed false information	0	0

and that whenever the speaker shares information in the truth of which they do not believe, the goal of the communication is simply to provide the hearer with options on the basis of which they could decide for themselves what the case actually is. The gains related to the perlocutionary goals are given in Table 2. The persuading goal is only fulfilled when the hearer accepts the believe, but the alternative goal is fulfilled by the simple act of telling, and the reaction of the hearer is irrelevant. I assume that the commitment messages (the basic message and the firsthand information message) are associated with the persuasion goal, while the hearsay message is chosen when the speaker aims only at presenting the hearer with options.

Table 2. Perlocutionary gains for the speaker

	Perlocutionary goal	
	Persuading	Presenting options
Transferred belief	10	3
Not transferred belief	0	3

Each message has an utterance cost and a conditional reputation cost, as shown in Table 3. The latter is only paid if the information turns out to be false. In case of sharing true information, there is a reputation *gain*. This aims at representing how one's utterances – according to their truth – contribute positively or negatively to one's reputation in the community.

Utterance costs are fixed, while the values of the reputation costs and gains are specified by the user (in the interval between 0 and 100). The chosen value for the reputation costs and gains is not only used to calculate the utility of the communication, but is also added to (or substracted from) the reputation of the speaker (which also varies between 0 and 100 and is initially 50).

Table 3. Costs and additional gains

	Utterance cost	Reputation cost	Reputation gain
Basic message (m_1)	2	[0, 100]	[0, 100]
Firsthand message (m_2)	3	[0, 100]	[0, 100]
Hearsay message (m_3)	3	[0, 100]	[0, 100]

The utility function for the speaker may thus be defined as follows:

$$U_s(m_i(e_j), a_k) = \begin{cases} O(m_i(e_j), a_k) + G_p(m_i, a_k) - C_u(m_i) + G_r(m_i), \\ \qquad\qquad\qquad\qquad \text{if } e_j \text{ happened} \\ O(m_i(e_j), a_k) + G_p(m_i, a_k) - C_u(m_i) - C_r(m_i), \\ \qquad\qquad\qquad\qquad \text{otherwise.} \end{cases} \quad (1)$$

where O refers to the basic outcome, C_u and C_r to the utterance and reputation costs, and G_r and G_p to the reputation and perlocutionary gains. $m_i(e_j)$ represents the uttering of a message of type m_i about event e_j. a_k for $k \in \{b, d\}$ is the action the hearer undertakes – either to believe (a_b) or doubt (a_d) the statement.

The utility function for the hearer is considerably simpler, as it equals the basic outcome:

$$U_h(m_i(e_j), a_k) = O(m_i(e_j), a_k). \quad (2)$$

The ultimate values of the utility functions of both players can be found in the game tree (Fig. 1), where the first number represents the expected utility for Player 1 (the speaker) and the second one – for Player 2 (the hearer).

4   Learning Mechanism

4.1   Description

I have chosen to model players' strategies and learning mechanisms with Pólya urns, much in the spirit of (Mühlenbernd 2011, pp. 6–8) and of the already existing Signaling Game NetLogo simulation (Wilensky 2016). Each player is modelled as having a set of speaker urns for their local speaker strategies and a set of hearer urns for their local hearer strategies.[12]

There are three urns for the three speaker information sets: Ω_w for when a witness, Ω_b for when heard and believed a report and Ω_n for when the report

---

[12] I call *local strategy* the strategy to act in a particular way if the game has already evolved to the state in which the player has to move. Simply *strategy* will refer to a combination of local strategies and will tell us how the player would move at any point of the game.

was *not* believed. Each urn contains two kinds of balls, but these two kinds are different for each information set. In Ω_w there are balls for the basic and the firsthand message: m_1 and m_2 respectively. In Ω_b there are balls for the basic and the hearsay message: m_1 and m_3. And in Ω_n there are balls for the hearsay message and for silence: m_3 and $m_\emptyset$.

There are other three urns for the three hearer information sets: Ω_{m_1} for the basic message, Ω_{m_2} for the firsthand information message and Ω_{m_3} for the hearsay information message. Each hearer strategy urn contains two kinds of balls: for believing the message (a_b) or for discarding it (a_d).

The number of balls b in an urn Ω at a time τ is written as $b(\Omega)_\tau$. At the beginning of the game, when $\tau = 0$, each player's urns have the content specified in Tables 4 and 5.

Table 4. The initial state of the urns for the speaker strategies

	$m_1(\Omega_s)_0$	$m_2(\Omega_s)_0$	$m_3(\Omega_s)_0$	$m_\emptyset(\Omega_i)_0$
$\Omega_s = \Omega_w$	100	100	0	0
$\Omega_s = \Omega_b$	100	0	100	0
$\Omega_s = \Omega_n$	0	0	100	100

Table 5. The initial state of the urns for the hearer strategies

	$a_b(\Omega_h)_0$	$a_d(\Omega_h)_0$
$\Omega_h = \Omega_{m_1}$	100	100
$\Omega_h = \Omega_{m_2}$	100	100
$\Omega_h = \Omega_{m_3}$	100	100

After each iteration of the game, the following strategy update is made for each speaker of type t, who utters a message m (or in other words – who drew a ball b_m from the urn Ω_t at time τ to report the event e) depending on the subsequent action a of the hearer:

$$m(\Omega_t)_{\tau+1} = \max[m(\Omega_t)_\tau + U_s(m(e), a), 1]. \tag{3}$$

Analogously, the strategy update for a hearer having drawn a ball b_a from urn Ω_m, i.e. who reacted with a to the utterance $m(e)$, would be:

$$a(\Omega_m)_{\tau+1} = \max[a(\Omega_m)_\tau + U_h(m(e), a), 1]. \tag{4}$$

The urn cannot contain less than one ball of each type, that has been allowed in it at the beginning of the game. In this way there is always a chance for the player to change their strategy.

4.2   Example

In order to make the calculation procedure clearer, I will go though one example here. Let's consider a pair of agents – a speaker s and a hearer h – at the beginning of the simulation. Let the speaker have firsthand evidence which leads them to forming a correct believe. Then let the speaker choose the marked message m_2 to communicate this believe and, finally, let the hearer adopt the believe, i.e. react with action a_b. In Fig. 1 this would be the uppermost branch. Let's assume further that the reputation cost/gain values are set to 20 and 10 for the firsthand and the basic message respectively, as it is assumed in Fig. 1.

Since the speaker has firsthand evidence, the update will be made on their urn for witnessed events Ω_w. The initial state of this urn is $[100, 100, 0, 0]$, which means equal inclination to use the firsthand information message or the unmarked message, and no option for choosing the hearsay message or silence. The update will concern the second value in the list, which is the number of the balls representing the inclination towards the firsthand message m_2. The update should be made according to Eq. 3 by providing the number of balls in the urn at time $\tau = 0$ and then calculating the speaker utility $U_s(m_2(e), a)$ for the move played.

The initial number of m_2 balls in Ω_w is given in Table 4. The speaker utility should be calculated according to the first case of Eq. 1, as we accept in this example that the speaker's belief is justified, i.e. the event e really happened:

$$U_s(m_2(e), a_b) = O(m_2(e), a_b) + G_p(m_2, a_b) - C_u(m_2) + G_r(m_2). \qquad (5)$$

$O(m_2(e), a)$ is the basic outcome for reporting an event e with a firsthand message m_2 to a speaker, who believes the message (i.e. reacts with action a_b). In this example the basic outcome is 10, as determined by the first row in Table 1.

The perlocutionary gain $G_p(m_2, a_b)$ is again 10, as specified in the first cell of the first column of Table 2 for the case of transferred belief with commitment messages.

The utterance cost $C_u(m_2)$ is 3, since m_2 is the firsthand message (see Table 3). And the reputation gain $G_r(m_2)$ is 20, as agreed for this example.

Once we make the substitutions in Eq. 1, we end up with the following calculation:

$$U_s(m_2(e), a_b) = 10 + 10 - 3 + 20 = 37. \qquad (6)$$

This is the value with which we have to update the number of the firsthand message balls in the firsthand information speaker urn. So, going back to Eq. 3 and substituting the variables with the corresponding values, we obtain:

$$m_2(\Omega_w)_1 = \max[100 + 37), 1] = 137. \qquad (7)$$

The speaker has a stronger inclination to choose the firsthand message m_2 after this successful run in the game.

For the hearer the procedure is much simpler. The updates will be made in the urn for the firsthand message Ω_{m_2} and more specifically, the number of the balls a_b for the hearer's inclination to believe this message will be updated. The initial number of these balls is 100 (see Table 5) and the only factor for the hearer utility is the basic outcome (see Eq. 2), which is 10 as is evident from Table 1. Therefore the calculation should be conducted as follows:

$$a_b(\Omega_{m_2})_1 = \max[a_b(\Omega_{m_2})_0 + U_h(m_2(e), a_b), 1] = \max[100 + 10, 1] = 110. \quad (8)$$

5   Visualization

The simulation is written in the language NetLogo,[13] and the explanation of its visualization will follow the structure of a typical NetLogo program. The basic element are the turtles,[14] these are the agents I use to represent communicating individuals. Then there are links between turtles – I represent by them the messages exchanged between the individuals.

5.1   Turtles

The turtles have shape, size, color and opacity. The shape represents the type of information source – the witnesses are square-shaped and the rest of the turtles have the shape of a circle. The size of the turtle represents the individual's reputation. The bigger the turtle, the greater its reputation.

Speaker local strategies are represented by color. The user can choose if they want to see the speaker local strategies for firsthand evidence, the ones for believed hearsay evidence or the ones for doubted hearsay evidence. In each case the probabilities of the individuals to use the three available messages (basic, firsthand information and hearsay information message) are mapped to the RGB color space. Red represents inclination towards the basic message, green – towards the firsthand information message and blue – towards the hearsay information message.

Opacity codes hearer local strategies. The user may choose the message for which to see the hearer local strategies. The turtles get the more opaque the more the individuals are inclined to believe the message. As simultaneous visualization of speaker and hearer local strategies may produce confusion, each of these visualizations can be disabled.

5.2   Links

The messages exchanged between individuals are represented with links between turtles. Color encodes type of message: red for basic message, green for firsthand

---

[13] See (Wilensky 1999).
[14] The language has been developed for simulating the behaviour of a robotized turtle, hence the extravagant name of this basic kind of agents.

information message and blue for hearsay information message. The color coding of links can be switched off.

The link is represented by a solid line if the transmitted information is true. Otherwise the line is dotted. If the hearer has believed the message, the line is completely opaque, otherwise its opacity is reduced.

6 Examples

Figure 2 presents the initial state of the game in a population of 100 individuals. The figure consists of three NetLogo views, representing the strategies for firsthand information, believed hearsay information and not believed hearsay information (in this order). As in the beginning it is equally likely for each agent to use the marked or the unmarked message to convey information based on firsthand evidence, the first view shows all turtles colored in equal proportions green (the color linked to the preference for the marked message) and red (coding the preference for the unmarked message). For the same reason in the second view the turtles are colored in violet – the blue in the mix represents the preference for the marked message for reported information, and the red – the preference for the unmarked message. The third view presents all turtles in half intense blue color, since there the two possible options are the marked message (connected to the blue color) and silence (full silence results in black coloring).

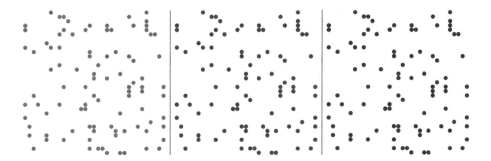

Fig. 2. Speaker local strategies for firsthand information, believed hearsay information and not believed hearsay information in the beginning of the simulation. (Color figure online)

Figures 3, 4 and 5 present the speaker strategies after 1000 communication 'steps' in a population of 100 individuals with 1 witness and reliability value of 0.9. What varies, are the values of the reputation costs/gains for the commitment messages (the basic and the firsthand message). Again, each figure consists of three NetLogo views, representing the strategies for firsthand information, believed hearsay information and not believed hearsay information.

Figure 3 presents the case in which the reputation of the agents is not influenced at all by what they say and the way they say it. This is why all the dots are the same size – the agents kept their initial reputation. There seems to be a slight preference for the marked message in the firsthand information scenario (most turtles are rather greenish) and somewhat clearer preference for the unmarked message in the believed hearsay case (there are many red turtles and only a few blue ones).

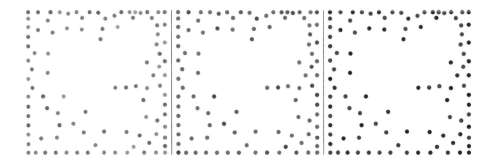

Fig. 3. Speaker local strategies for firsthand information, believed hearsay information and not believed hearsay information, with no impact of the commitment messages on the speaker's reputation. (Color figure online)

Figure 4 is an example for the influence of a high reputation bet value (80 for both commitment messages). One can see how the dots are of different sizes, representing agents with different reputation levels. Furthermore, there is a clear tendency for marking hearsay information. The agents seem to have divided in their strategies towards firsthand information. In comparison with Fig. 3, here the colors are brighter, which means that the speakers' preferences are clearer. These clear preferences are common for the community in the case of hearsay and more a matter of personal choice in the firsthand scenario.

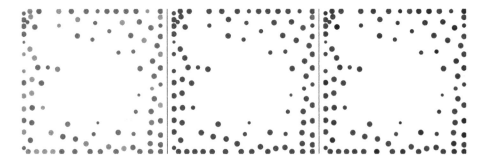

Fig. 4. Speaker local strategies for firsthand information, believed hearsay information and not believed hearsay information, with high impact of the commitment messages (reputation bet = 80) on the speaker's reputation. (Color figure online)

Figure 5 consists of two parts: Case A and Case B. They are two different developments that occur when the simulation is run twice with the same initial parameters, viz. reputation bet value of 80 for the firsthand message and 60 for the basic.

Case A.

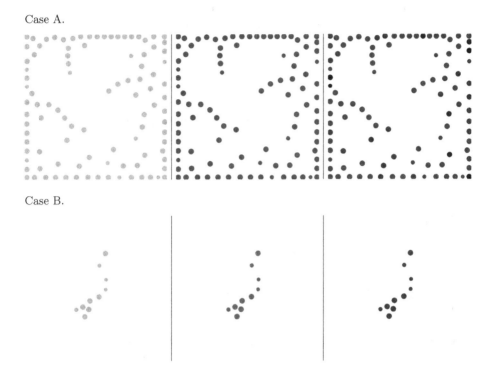

Case B.

Fig. 5. Speaker local strategies for firsthand information, believed hearsay information and not believed hearsay information, with different impacts of the commiting messages (reputation bet for firsthand message = 80, and for basic message = 60) on the speaker's reputation. Cases A. and B. are different developments of the same initial settings. (Color figure online)

In Case A the whole population managed to develop a strategy to mark hearsay information, as well as firsthand information. In Case B the population again developed a preference (somewhat weaker, though) for marking firsthand information, but this time they failed to adopt a strategy for marking hearsay. As a result, it is more likely for an agent in Case B to loose reputation and ultimately be excluded from the community, which is why there are so few agents remaining in case B, even though their initial number was 100, like in Case A.

7   Conclusion

I have described here a simulation that presents speaker reputation as one of the factors relevant for the systematic marking of information source. It was shown that the impact of one's speech on their reputation does influence the choice to indicate the information source or not. Furthermore, we saw that in a setting with high impact of speech on reputation *not* marking hearsay information increases the risk of exclusion from the community.

The finding that reputation and systematic marking of information source are related can explain (at least to some extent) the existence of the grammatical category evidentiality in some linguistic communities. It is in line with the fact that most languages with large evidential systems are spoken in small, compact communities, where a person is very dependent on their good name.

Acknowledgments. I would like to thank Dessislava Ivanova for bringing NetLogo to my attention, Hristo Todorov and Evgeni Latinov for their comments on a number of drafts of this paper, and the anonymous reviewers for their feedback and suggestions.

References

Aikhenvald, A.Y.: Evidentiality. Oxford University Press, Oxford (2004)

Faller, M.: Evidentiality and epistemic modality at the semantics/pragmatics interface (2006). https://www.academia.edu/25944467/Evidentiality_and_Epistemic_Modality_at_the_Semantics_Pragmatics_Interface

Mühlenbernd, R.: Learning with neighbours. Synthese **S1**(183), 87–109 (2011)

Smirnova, A.: The meaning of the Bulgarian evidential and why it cannot express inferences about the future. Proc. SALT **21**, 275–294 (2011)

Wilensky, U.: NetLogo. Center for connected learning and computer-based modeling, Northwestern University, Evanston, IL (1999). http://ccl.northwestern.edu/netlogo/

Wilensky, U.: NetLogo Signaling Game model. Center for connected learning and computer-based modeling, Northwestern University, Evanston, IL (2016). http://ccl.northwestern.edu/netlogo/models/SignalingGame

Disjunction Under Deontic Modals: Experimental Data

Ying Liu[✉]

Utrecht University, Utrecht, Netherlands
y.liu9@uu.nl

Abstract. The meaning components of *may/or* and *must/or* sentences have been discussed intensively by a number of theoretical accounts. The debates are concerned with whether free choice inferences are part of logical meaning or scalar implicatures, and additionally, whether exhaustive inferences and exclusive *or* inferences are derived for *may/or* versus *must/or*. In this study, two experiments were conducted for the purpose of evaluating the assumptions of three representative accounts, namely, Fox [10], Geurts [12] and Simons [20]. Each experiment separately examined the availability and processing time-course of the three types of inferences associated with *may/or* versus *must/or* sentences. The experimental results fit Simons's [20] analysis to a large extent.

1 Introduction

This experimental study has two focuses. First, it presents a set of contrastive data that illustrates how people interpret and process *may/or* sentences, i.e., sentences with disjunction embedded under a deontic possibility modal as shown in (1), versus *must/or* sentences, i.e., sentences with disjunction embedded under a deontic necessity modal as shown in (2). Second, it discusses to what extent different theoretical accounts explain the data by looking into their assumptions.

(1) Mary may eat an apple or a banana.   ***may/or*** sentence

(2) Mary must eat an apple or a banana.   ***must/or*** sentence

Intuitively, both (1) and (2) yield a strong free choice inference as shown in (3). This inference is special in the sense that the standard semantics (i.e., the combination of the Boolean analysis of disjunction [2] and the standard modal logic for deontic modals [16]) completely fails to explain it, while the standard Neo-Gricean reasoning (e.g. [19]) can only account for its derivation for *must/or* sentences.

My thanks go to Dr. Yaron McNabb for supervising the study, Dr. Rick Nouwen and Dr. Henriette de Swart for suggestions, Chris van Run for the ZEP scripts and the anonymous reviewers for the feedback.

J. Sikos and E. Pacuit (Eds.): ESSLLI 2018, LNCS 11667, pp. 185–199, 2019.
https://doi.org/10.1007/978-3-662-59620-3_12

(3) free choice inference (options indicated by each disjunct are permitted):
    Mary is permitted to eat an apple, and she is also permitted to eat a banana.

In order to universally explain free choice inferences, in general, two types of accounts are developed: scalar implicature accounts [3,10,11,13,22], and semantic accounts [1,4,5,12,14,20,21,24]. The main divergence of those accounts lies in how to explain the nature of free choice inferences. More specifically, scalar implicature accounts assume that free choice inferences can be derived under the same mechanism as that for deriving scalar implicatures, while semantic accounts suggest that free choice inferences are part of logical meaning, and are derived as the result of the computation of truth conditions.

In addition to free choice inferences, the interpretations of (1) and (2) are also closely associated with the other two types of inferences: the exhaustive inference as shown in (4), and the exclusive or inference as shown as (5).

(4) exhaustive inference (options not indicated by disjuncts are not permitted):
    Mary is not permitted to eat anything other than an apple or a banana.
(5) exclusive or inference (only one option is permitted at a time):
    Mary is not permitted to eat both an apple and a banana.

With regard to the puzzle whether exhaustive and exclusive or inferences are derivable for may/or and must/or sentences, different theoretical accounts have different assumptions. There are accounts, such as [12], which assume that exhaustive and exclusive or inferences are similarly available for both may/or and must/or sentences. However, there are also accounts, such as [3,10,20], which suggest that there exists a contrast between may/or and must/or sentences in the degree of availability of either exhaustive inferences or exclusive or inferences.[1]

Previous experimental studies (e.g. [7,8,23]) limit their investigations to free choice inferences drawn from may/or sentences. But none of them looks into the availability of exhaustive inferences or exclusive or inferences associated with may/or sentences. And there is also none of them that examines any inference associated with must/or sentences. Therefore, in this study, I carried out two experiments for the purpose of answering the following questions: first, whether and to what extent are free choice, exhaustive and exclusive or inferences available for may/or and must/or sentences? Second, what is the derivation mechanism of the three types of inferences? Third, can any difference be observed between may/or and must/or sentences? The paper is structured as follows. In Sect. 2, I briefly review the assumptions of three representative accounts, namely, Fox [10], Geurts [12] and Simons [20], and discuss how those assumptions are

---

[1] More specifically, based on Simons's [20] account (i.e., [20]), the contrast between may/or and must/or sentences lies in the availability of exhaustive inferences, while based on Fox [10] or Alonso-Ovalle's [3] account (i.e., [10] or [3]), the contrast between the two sentences lies in exclusive or inferences instead.

experimentally testable.[2] In Sects. 3 and 4, I separately report the experimental investigations on *may/or* and *must/or* sentences. In Sect. 5, I compare the data of *may/or* sentences with those of *must/or* sentences, and discuss how the data may shed light on future theories.

2   Background

Scalar Implicature Account. Inspired by Kratzer and Shimoyama [17] and Alonso-Ovalle [3], Fox [10] developed an algorithm to compute free choice inferences as a special kind of scalar implicatures. Following Kratzer and Shimoyama, Fox suggests that the foundation for deriving free choices lies in what is known as the "anti-exhaustivity" principle (p. 13). To put it briefly, different from a standard scalar implicature that can be derived by negating the alternatives of its corresponding assertion, free choice inferences are assumed to be derived for sentences such as (1) by negating the strengthened meanings of the alternatives of those sentences. Under Fox's [10] account, anti-exhaustivity can be realized through the recursive parsing of the sentences, by using a covert exhaustive operator (Exh) with a meaning similar to *only*.

(6) a. Exh(C)($\diamond$(Mary eat an apple or a banana))   first parsing
       C = {$\diamond$(Mary eat an apple or a banana), $\diamond$(Mary eat an apple),
              $\diamond$(Mary eat a banana), $\diamond$(Mary eat an apple and a banana)}
    b. Exh(C')[Exh(C)($\diamond$(Mary eat an apple or a banana))]   second parsing
       C' = {$\diamond$(Mary eat an apple or a banana)$\wedge \neg \diamond$(Mary eat an apple and
              a banana), $\diamond$(Mary eat an apple but not a banana),$\diamond$(Mary eat a
              banana but not an apple)}
                                    (see details in Fox [10], pp. 30–32)

Exh needs to operate over a set of alternatives. The set of alternatives of a *may/or* or *must/or* sentences can be generated by replacing the corresponding disjunction with its logically non-weaker scale-mates within the same Horn-scale (Fox follows Sauerland's [19] definition of the Horn-scale for disjunction). The alternatives of (1) are listed in the set C in (6a). Note that Fox suggests that Sauerland's alternatives can be understood as "basic answers to a Hamblin-question closed under disjunction" (footnote 29, p. 19). This is to say, if the question relevant to (1) is "what is Mary allowed to eat?", then the only possible answers are "an apple" and "a banana". Since all other food options are irrelevant and impossible answers, they are therefore excluded from the set of alternatives from the very beginning. The same reasoning can also be applied to (2). In this sense, exhaustive inferences such as (4) would always be present for both *may/or* and *must/or* sentences.

---

[2] I selected these accounts for discussion because they come up with a uniform analysis for both *may/or* and *must/or* sentences and they make assumptions about all three types of inferences. Many accounts in literature, e.g. Barker [5] and Starr [21], mention nothing about *must/or* sentences.

One primary role of Exh is to eliminate as many incorrect ignorance inferences as possible. When Exh is applied to (1) at the matrix position for the first time as shown in (6a), the only alternative that can be negated is $\Diamond$(Mary eat an apple and a banana), and the negation of this alternative gives rise to the exclusive *or* inference in (5).[3] After the first parsing, an incorrect ignorance inference which implies that the speaker does not know which of an apple and a banana Mary is allowed to eat is also generated. This inference triggers the second parsing as shown in (6b). In the second parsing, Exh operates over the strengthened meanings of the alternatives that are not negated in the first parsing in the set C in (6a). Those strengthened alternatives are listed in the set C' in (6b). By simultaneously negating $\Diamond$(Mary eat an apple but not a banana) and $\Diamond$(Mary eat a banana but not an apple), the corresponding free choice inference given in (3) can be generated.

(7) Exh(C)($\Box$(Mary eat an apple or a banana))
    C = {$\Box$(Mary eat an apple or a banana), $\Box$(Mary eat an apple), $\Box$(Mary eat a banana), $\Box$(Mary eat an apple and a banana)}

Different from *may/or* sentences, a *must/or* sentence such as (2) only involves a one-step parsing at the matrix position as shown in (7). All alternatives listed in the set C in (7) except (2) itself can be negated simultaneously without contradicting the truth of (2). As a result, the inference that the speaker believes that $\neg$ $\Box$(Mary eat an apple) and $\neg$ $\Box$(Mary eat a banana) can be derived.[4] This inference is compatible with the free choice inference in (3). Note that here the exclusive *or* inference is not derivable. The reason is that by negating $\Box$(Mary eat an apple and a banana), the inference that implies that it is not obligatory for Mary to eat both an apple and a banana is derived. This inference is compatible with the type of world in which Mary is permitted to eat both an apple and a banana. It conveys a meaning that is opposite from the desired exclusive *or* inference in (5).

Semantic Accounts. Geurts [12] and Simons [20] solve the free choice puzzle by proposing novel semantic analyses for disjunction or/and modals. Their accounts are crucially different from each other in how they deal with the scope relation between disjunction and deontic modals and how they formulate the semantics for disjunction.

More specifically, Geurts claims that if one assumes that disjunction can take scope over deontic modals (following Zimmermann [24]), then sentences with disjunction embedded under deontic modals can be analyzed as conjunctions of

---

[3] The reason why the stronger alternatives containing the individual disjuncts, i.e., $\Diamond$(Mary eat an apple) and $\Diamond$(Mary eat a banana), cannot be negated is that they are not "innocently excludable" (p. 24). If they are negated at the same time, the derived inferences will be in conflict with the truth of (1).

[4] $\neg$ $\Box$(Mary eat an apple) is equivalent to $\Diamond$ $\neg$(Mary eat an apple), but is crucially different from $\neg$ $\Diamond$ (Mary eat an apple). So the inference that Mary is not under the obligation to eat an apple does not imply that Mary is not allowed to eat an apple.

modal propositions.[5] This is to say, the logical forms of the *may/or* sentence in (1) and the *must/or* sentence in (2) can be written as (8) and (9) respectively.

(8) $\Diamond_{D1}(\lambda_w.\ \text{eat}(\text{Mary, an\_apple}))\bigwedge \Diamond_{D2}(\lambda_w.\ \text{eat}(\text{Mary, a\_banana}))$
    where $D_1 = D_2 = \text{ACC}_{deontic,w}.$
(9) $\Box_{D1}(\lambda_w.\ \text{eat}(\text{Mary, an\_apple}))\bigwedge \Box_{D2}(\lambda_w.\ \text{eat}(\text{Mary, a\_banana}))$
    where $D_1 \subseteq \text{ACC}_{deontic,w}.$ and $D_2 \subseteq \text{ACC}_{deontic,w}.$

In order to correctly derive free choice inferences, Geurts suggests that by default the domains of quantification of the two deontic possibility modals in (8) are identical to each other, and they are also identical to the set of deontically accessible worlds. As a result, in the set of worlds of evaluation of (8), at least one world in which Mary eats an apple is deontically accessible, and similarly, at least one world in which Mary eats a banana is also deontically accessible. In comparison, the two deontic necessity modals in (9) have distinct domains of quantification (which might partly overlap), and each of their domains of quantification only partly covers the set of deontically accessible worlds. Therefore, the set of deontically accessible worlds of (9) is partitioned into two subsets. One subset contains the worlds in which Mary eats an apple, while the other subset contains the worlds in which Mary eats a banana. This is to say that the worlds in which Mary eats an apple, as well as the worlds in which Mary eats a banana, are possibly but not obligatorily accessible. Since in Geurts' analysis disjunction always takes wide scope in *may/or* and *must/or* sentences, the existence of the exhaustive effect is completely dependent on whether or not the list of modal propositions is closed under disjunction. Following Zimmermann, Geurts claims that unless sentences such as (1) and (2) are explicitly marked by intonation or other linguistic devices, a "exhaustivity" constraint should be put on them by default. This constraint functions to closes the conjunctive list of modal propositions coordinated by disjunction, and is assumed to be a "semantic" constraint (p. 385). Thus, based on Geurts's [12] account, exhaustive inferences are supposed to be usually derivable in semantics for both *may/or* and *must/or* sentences. Geurts also proposes a "disjointness" constraint for both *may/or* and *must/or* sentences. By applying this constraint, the overlap between the worlds indicated by different individual disjuncts can be prohibited, and exclusive *or* inferences are therefore generated. Geurts suggests that the disjointness constraint is "a conversational implicature of sorts, though it is not a scalar implicature" (p. 403).

By contrast, Simons [20] proposes that under the scope of deontic modals, disjunction introduces sets of alternative propositions by going through a process known as "independent composition" (p. 288). Independent composition is preferably halted at deontic modals due to certain unspecified reasons. (10) and (11) below illustrate the logical forms of (1) and (2) after independent composition.

---

[5] Geurts [12] adopted novel semantics for both disjunction and modals. More specifically, he followed Zimmermann's [24] idea to make "or" present a list of alternatives. And inspired by Kratzer [15], he further suggested that expressions containing modals are context-dependent and this context-dependence might be of a presuppositional nature.

(10) ◇ {{eat(Mary, an_apple)}, {eat(Mary, a_banana)}}
(11) □ {{eat(Mary, an_apple)}, {eat(Mary, a_banana)}}

Simons further proposes the idea of "supercover" condition of *or* (p. 276). To keep things simple, when *or* is in the "supercover" condition, it functions to divide up certain domains on the basis of the contents of the disjuncts. To specify, (10) is true if and only if there exists at least one subset in the set of deontically accessible worlds, which contains a type of world in which Mary eats an apple and a type of world in which Mary eats a banana. Note that the set of deontically accessible worlds of (10) may also contain other types of world, such as the type of world in which Mary eats a pear, the type of world in which Mary eats a cherry, etc., so (10) is not exhaustive. In comparison, (11) is true if and only if the set of deontically accessible worlds itself is categorized into two types of worlds: the type of world in which Mary eats an apple, and the type of world in which Mary eats a banana. Since the type of world in which Mary eats a thing other than an apple or a banana are inaccessible, the exhaustive inference is semantically generalized for (11). Similar as Geurts, Simons suggests that exclusive *or* inferences can be derived for both *may/or* and *must/or* sentences if some kind of pragmatic constraint, such as the "alternativeness" constraint (p. 31), is put on the sentences.

Psychological Implications. Studies on language processing (e.g. [6,9,18])[6] suggest that the computation of logical meaning takes place alongside with the computation of logical forms in syntax, so it is automatic with low cognitive costs involved. In comparison, scalar implicature derivation is cognitively costly, because hearers may need spend cognitive resources to retrieve the alternatives of a certain scalar expression from working memory, to decide whether the reasoning process involving the calculation of scalar implicatures should be activated, or to reason why speakers utter a specific scalar expression instead of its stronger alternatives. Since cognitive resources in working memory are limited, it is hard for people to derive scalar implicatures for every single occurrence of scalar expressions. As a result, scalar implicatures are only optionally derived.

In psycholinguistic experiments, the degree of availability of an inference, reflected by derivation rates, conveys information about whether this inference is computed automatically or optionally; while the processing time conveys information about whether an inference is cognitively costly. Based on these, if an inference is derived semantically, the derivation rate of it should be ideally near 100%, and the processing time of it should be similar as that of a logical interpretation. By contrast, if an inference is a scalar implicature in nature, the derivation rate of it should be moderate, i.e., a value that is neither close to 0% nor 100%, and the processing time of it should be much longer than that of a logical interpretation. Thus, by experimentally examining the derivation rate

---

[6] Considering the fact that in psycholinguistic experiments the difference in the types of tasks involved causes crucial differences in results, here I only discuss the experimental studies which adopt the experimental method similar as the one I used, i.e., a binary picture-sentence verification task.

and processing time, we can tell whether or not an inference is available and whether it is part of logical meaning or a scalar implicature.

Based on what is mentioned above, once we obtain the derivation rate and processing time of each of the three types of inferences associated with *may/or* and *must/or* sentences, we can immediately investigate the meaning components of *may/or* versus *must/or* sentences as well as the nature of these components. Subsequently, we can evaluate the plausibility of different theoretical accounts.

3   Experiment for *May/or* Sentences

Purposes. To examine derivation rates and processing time of free choice, exhaustive and exclusive *or* inferences triggered by *may/or* sentences.

Participants 40 Dutch native speakers (aged 18 and above) were recruited from the Dutch participant database of Utrecht University. The experiment took them around 30 min to finish. Each participant was compensated with 5 euros.

The picture shows a lottery machine for children. After a child wins a lottery game, the small screen on the right hand side of the machine will display the total amount of cash prize the child has been awarded. The central screen of the machine will display six different items. The price and availability of the items are displayed below each one of the items. The price is indicated in Euros. A green light indicates an available item and a red light indicates that the item is not available for purchase. The price of the items and their availability as well as the amount of cash prize the child has been awarded all determine what items the child is allowed to buy from the lottery machine.

Fig. 1. English translation of the cover story for *May/or* experiment

Method. I designed a "lottery game" paradigm in Dutch in ZEP [7], which required participants to do an online binary picture-sentence verification task based on the cover story shown in Fig. 1. The cover story was only presented once at the beginning of the task. Participants were asked to read the cover story carefully without a time constraint, and their understandings about the story were examined by 6 practice trials. Only when they successfully passed all practice trials, they could start the test session.

Figure 2 is an example that illustrates how target trials look like. For each trial, a picture depicting a lottery machine would first be presented on a computer screen for 500 ms. After this, a plus sign ("+") would occur at the beginning of the sentence bar beneath the picture. Participants were instructed that if they press the middle button on the button box in front of them, the plus sign would be replaced by the first chunk of the sentence. The sentence chunks could continuously show up by pressing the same button. Participants could read the sentence at their own pace by controlling the speed of button press. All *may/or* sentences were cut into five chunks as shown in Fig. 2. Once the entire sentence was presented on the screen, participants were required to judge whether the

---

[7] Information about ZEP can be found at https://www.beexy.nl/.

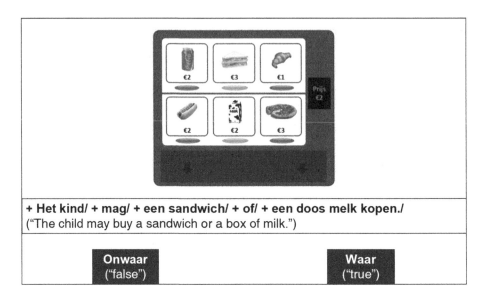

Fig. 2. Example of target trial in *May/or* experiment

sentence is true or false based on the cover story by pressing the corresponding left/right button on the button box. *True/false* responses and the reaction time from the occurrence of the last sentence chunk to the left/right button press were recorded.

The *false* response in Fig. 2 indicates the existence of the free choice inference. To illustrate, based on the cover story, the picture indicates that the only item the child is permitted to buy is a box of milk because it is both available and affordable. If participants derive the free choice inference for the sentence, which implies that the child is permitted to buy a sandwich and he/she is also permitted to buy a box of milk, they should judge the sentence as the incorrect description of the picture. Thus, the percentage of *false* responses in the trial shown in Fig. 2 indicates the derivation rates of free choice inferences, while the reaction time of these *false* responses reflects the processing time needed for deriving free choice inferences. By adopting a similar design, the data of exhaustive and exclusive *or* inferences were also obtained.

Design and Materials. A single factor within-subject design was used. The independent variable was the type of conditions created for *may/or* sentences. Each *may/or* sentence occurred once in three target conditions and two control conditions (see Fig. 3). There were two dependent variables: the type of responses (i.e., *true* or *false*) and the reaction time associated with the responses. 12 trials were created for each target or control condition. 70 filler trials were added to conceal experimental purposes. All trials were pseudo-randomized. The sentences used for creating filler trials are, for example, *some types of fruits are available, some accessories are not available, both a cucumber and a tomato cost one euro, the cheapest item is a lemon, the most expensive item is a scarf, no drink is available.*

Target Sentence
Het kind mag een sandwich of een doos melk kopen. ("The child may buy a sandwich or a box of milk.")

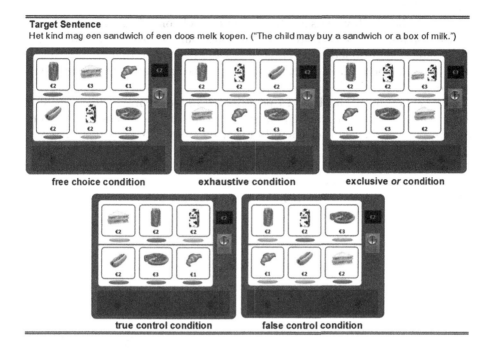

Fig. 3. Target and control conditions in *May/or* experiment

In total, each participant saw 136 trials. All trials were pseudo-randomized, so that no trial from the same condition occurred twice in a row.

Predictions. Fox [10] predicts that exhaustive inferences are derived semantically, while free choice and exclusive *or* inferences are derived as scalar implicatures. Based on this, the percentage of *false* responses in the exhaustive condition should be near 100%, and the corresponding reaction time should be roughly the same as that in control conditions. The percentage of *false* responses in the free choice and exclusive *or* condition should be neither close to 0% nor 100%, and the corresponding reaction time should be longer than that in control conditions.

Geurts [12] predicts that exhaustive and free choice inferences are derived semantically, while exclusive *or* inferences are derived as conversational implicatures. Based on this, the percentage of *false* responses in the free choice and exhaustive condition should be near 100%, and the corresponding reaction time should be roughly the same as that in control conditions. The percentage of *false* responses in the exclusive *or* condition should be neither close to 0% nor 100%, and the corresponding reaction time should be longer than that in control conditions.

Simons [20] predicts that free choice inferences are derived semantically and exclusive *or* inferences are derived as conversational implicatures, while exhaustive inferences are absent. Based on this, the percentage of *false* responses in the free choice condition should be near 100%, and the corresponding reaction time should be roughly the same as that in control conditions. The percentage of *false* responses in the exclusive *or* condition should be neither close to 0% nor 100%, and the corresponding reaction time should be longer than that in control conditions. The percentage of *false* responses in the exhaustive condition should be near 0%.

Results. I excluded the data of 7 participants whose accuracy rates on control items were below 60%. I removed all reaction time data associated with wrong responses in control conditions, and I further removed 5.78% reaction time data which are outliers[8].

I submitted the response data to a generalized linear mixed effects model in R. The model indicates that the difference in conditions had an significant influence on responses, $F(4) = 104.61$, $p < 0.001$.

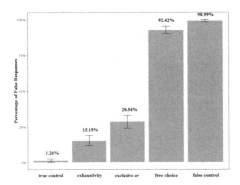
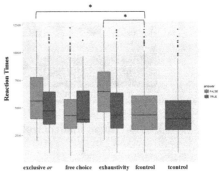

Fig. 4. *False* Responses (*May/or*)   **Fig. 5.** Reaction Time (*May/or*)

The percentages of *false* responses, indicating the derivation rates of the inferences under investigation, are given in Fig. 4. The derivation rate of free choice inferences is as high as 92.42%. The derivation rates of exhaustive and exclusive *or* inferences were much lower, i.e., 15.15% and 28.54% respectively. The percentage of *false* responses in the free choice condition was significantly higher than that in the true control condition ($\beta = 10.77$, SE $= 0.66$, z $= 16.19$, p < 0.001), and it was significantly lower than that in the false control condition ($\beta = -2.36$, SE $= 0.57$, z $= -4.13$, p < 0.001). The percentage of *false* responses in the exhaustivity condition was significantly higher than that in the true control condition ($\beta = 3.49$, SE $= 0.53$, z $= 6.60$, p < 0.001), and it was significantly lower than that in the false control condition ($\beta = -9.63$, SE $= 0.68$, z $= -14.16$, p < 0.001). The percentage of *false* responses in the exclusive *or* condition was significantly higher than that in the true control condition ($\beta = 5.23$, SE $= 0.56$, z $= 9.35$, p < 0.001), and it was significantly lower than that in the false control condition ($\beta = -7.90$, SE $= 0.63$, z $= -12.50$, p < 0.001).

---

[8] Outliers are more than 1.5 IQRs below the first quartile or above the third quartile.

In addition, significant differences in the percentage of *false* responses were detected across all three target condition. The percentage of *false* responses in the free choice condition was significantly higher than that in the exhaustive condition (β = 7.28, SE = 0.45, z = 16.22, p < 0.00) and that in exclusive *or* condition (β = 5.54, SE = 0.37, z = 14.86, p < 0.00). The percentage of *false* responses in the exhaustive condition was significantly lower than that in the exclusive *or* condition (β = −1.74, SE = 0.29, z = −6.00, p < 0.00) and that in the free choice condition (β = −7.28, SE = 0.45, z = −16.22, p < 0.00). The percentage of *false* responses in the exclusive *or* condition was significantly higher than that in the exhaustive condition (β = 1.74, SE = 0.29, z = 6.00, p < 0.00), but it was significantly lower than that in the free choice condition (β = −5.54, SE = 0.37, z = −14.86, p < 0.00).

I submitted the reaction time data to a linear mixed effects model in R. The model indicates that the difference in conditions and responses had an significant influence on reaction time, $F(7) = 11.09$, p < 0.001.

The reaction time, reflecting the processing time of the inferences under investigation, is illustrated in Fig. 5. The reaction time of free choice inferences (M ≈ 4724 ms, SD ≈ 2228 ms) were not significantly different from that in the true and false control condition. The reaction time associated with exhaustive inferences (M ≈ 6551 ms, SD ≈ 2367 ms) was significantly longer than that in the false control condition (β = 9.60, SE = 2.04, t = 4.68, p < 0.001) and the true control condition (β = 11.8, SE = 2.04, t = 5.78, p < 0.001). The reaction time associated with exclusive *or* inferences (M ≈ 5968 ms, SD ≈ 2680 ms) was significantly longer than that in the false control condition (β = 7.10, SE = 1.61, t = 4.42, p < 0.001) and the true control condition (β = 9.30, SE = 1.60, t = 5.81, p < 0.001). In addition, the reaction time associated with free choice inferences was significantly shorter than that associated with exclusive *or* inferences (β = −7.20, SE = 1.61, t = −4.49, p < 0.001) and the exhaustive inferences (β = −9.70, SE = 2.05, t = −4.74, p < 0.001). The reaction time associated with the exclusive *or* inferences was slightly shorter than that associated with the exhaustive inferences, but this difference was not significant (β = −2.50, SE = 2.28, t = −1.08, p ≈ 0.28).

Discussion. There are three important findings. First, free choice inferences were derived almost by default, and the derivation of them was not more time-consuming than that of logical meanings. The similar finding was also reported in Chemla and Bott [8]. Second, both exhaustive and exclusive *or* inferences were only optionally derived, with the processing time significantly longer than that of logical interpretations. Third, the derivation rate of free choice inferences was much higher than that of exhaustive and exclusive *or* inferences; while the processing time of free choice inferences was significantly shorter than that of exhaustive and exclusive *or* inferences. The findings suggest that the cognitive behaviors of exhaustive and exclusive *or* inferences are more or less similar as each other, while the cognitive behaviors of free choice inferences are crucially different from both exhaustive and exclusive *or* inferences.

Based on the psychological implications reviewed in Sect. 2, I argue that free choice inferences are most likely to be the preferred logical interpretations of *may/or* sentences, while exhaustive and exclusive *or* inferences are most likely to be some sort of conversational implicatures if they indeed are not scalar implicatures. Fox [10] account has the poorest fit to the data because it completely fails to predict the derivation pattern of free choice and exhaustive inferences. Geurts [12] faces a fatal problem in explaining the extremely low derivation rate of exhaustive inferences. Loosely speaking, Simons's [20] account has a considerably good fit to the data. It can coherently explain the rapid and near-default derivation of free choice inferences, as well as the optional and delayed derivation of exclusive *or* inferences. Furthermore, it is the only account among the three, which can reasonably explain why exhaustive inferences were largely absent in the interpretation of *may/or* sentences.

4   Experiment for *Must/or* Sentences

Purposes. To examine derivation rates and processing time of free choice, exhaustive and exclusive *or* inferences triggered by *must/or* sentences.

Participants. 25 Dutch native speakers (aged 18 and above) were recruited from the Dutch participant database of Utrecht University. The experiment took them around 30 min to finish. Each participant was compensated with 5 euros.

Methods, design and materials The paradigm, set-up, design and materials used in this experiment were exactly the same as those of the *may/or* experiment (see Sect. 3) except for two aspects. First, one piece of information was added to the cover story of this experiment, which stated that the child has to buy something from the lottery machine with the cash prize he/she has been awarded, otherwise the machine will be unable to load the next lottery game. Second, *moeten* (*"must"*) instead of *mogen* (*"may"*) was used in all target sentences.

Predictions. Fox [10] predicts that exhaustive inferences are derived semantically and free choice inferences are derived as scalar implicatures, while exclusive *or* inferences are not derived. Based on this, the percentage of *false* responses in the exhaustive condition should be near 100%, and the corresponding reaction time should be roughly the same as that in control conditions. The percentage of *false* responses in the free choice condition should be should be neither close to 0% nor 100%, and the corresponding reaction time should be longer than that in control conditions. The percentage of *false* responses in the exclusive *or* condition should be near 0%.

Geurts [12] and Simons [20] predict that exhaustive and free choice inferences are derived semantically, while exclusive *or* inferences are derived as conversational implicatures. Based on this, the percentage of false responses in the free choice and exhaustive condition should be near 100%, and the corresponding reaction time should be roughly the same as that in control conditions. The percentage of *false* responses in the exclusive *or* condition should be neither close to 0% nor 100%, and the corresponding reaction time should be longer than that in control conditions. Note that Geurts's [12] predictions for *may/or* and *must/or* sentences are exactly the same.

Results. I included all participants' response data. I removed all reaction time data associated with wrong responses in control conditions, and I further removed 5.47% reaction time data which are outliers.

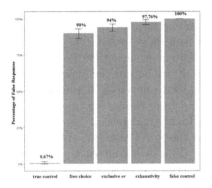

Fig. 6. *False* responses (*Must/or*)

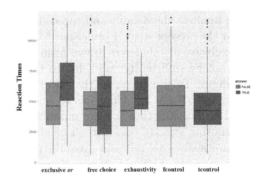

Fig. 7. Reaction time (*Must/or*)

According to Fig. 6, the derivation rates of free choice, exhaustive and exclusive *or* inferences associated with *must/or* sentences were all very high, i.e., 90%, 97.76% and 94% respectively. The statistical results suggest that the difference in conditions had a significant influence on responses, $F(4) = 23.87$, $p < 0.001$. The percentages of *false* responses in all three target conditions were only significantly different from that in the true control condition, but they were not significantly different from that in the false control condition. More specifically, the percentage of *false* responses in the false control condition was not significantly higher than that in the free choice condition ($\beta = 19.17$, $SE = 2284.02$, $z = 0.01$, $p \approx 0.99$), the exclusive or condition ($\beta = 18.45$, $SE = 2284.02$, $z = 0.01$, $p \approx 0.99$), and the exhaustivity condition (($\beta = 17.15$, $SE = 2284.02$, $z = 0.01$, $p \approx 0.99$).

Figure 7 summarizes the reaction time data associated with *must/or* sentences. The reaction time of free choice inferences ($M \approx 4715$ ms, $SD \approx 2387$ ms) were not significantly different from that of the false control condition ($\beta = -0.40$, $SE = 1.31$, $t = -0.32$, $p \approx 0.75$) and the true control condition ($\beta = 1.10$, $SE = 1.31$, $t = 0.87$, $p \approx 0.39$). The reaction time of exhaustive inferences ($M \approx 4659$ ms, $SD \approx 2326$ ms) were not significantly different from that of the false control condition ($\beta = -0.40$, $SE = 1.28$, $t = -0.34$, $p \approx 0.74$) and the true control condition ($\beta = 1.10$, $SE = 1.28$, $t = 0.88$, $p \approx 0.38$). The reaction time of exclusive *or* inferences ($M \approx 5013$ ms, $SD \approx 2594$ ms) were also not significantly different from that of the false control condition ($\beta = 1.70$, $SE = 1.30$, $t = 1.28$, $p \approx 0.21$) and the true control condition ($\beta = 3.20$, $SE = 1.30$, $t = 2.48$, $p \approx 0.02$).

Discussion. The main finding is that all three types of inferences were found to be derived by default for *must/or* sentences, with the processing time not significantly different from that of logical interpretations in control conditions.

Therefore, it seems that all three types of inferences are parts of the preferred logical interpretations of *must/or* sentences. Fox's [10] account still only very poorly fits the data. It only predicts the derivation pattern of exhaustive inferences. Both Geurts's [12] and Simons's [20] accounts successfully explain the data associated with free choice and exhaustive inferences; however, none of them predicts the default and rapid derivation of exclusive *or* inferences.

At the present stage, I prefer not to make any firm claim towards the nature of the exclusive *or* inferences drawn from *must/or* sentences, since the relevant data may not be sufficiently valid due to the potential problem in the experimental paradigm. To illustrate, in the cover story, I added one piece of information which stated that the child has to buy something from the lottery machine. It could be possible that a large number of participants understood "something" as denoting exactly one instead of at least one thing. As a result, they might automatically rule out the possibility that the child is permitted to buy two things at once. Thus, More experiments need be done to better understand the nature of exclusive *or* inferences triggered by *must/or* sentences.

5   Conclusion

If comparing the data of *may/or* with those of *must/or* sentences, one similarity and two differences can be found. The similarity is that free choice inferences are the preferred logical interpretations of both types of sentences. So generally, the semantic accounts, i.e., Geurts [12] and Simons [20], are more plausible than the scalar implicature account, i.e., Fox [10].

The crucial contrasts between the data associated with *may/or* and *must/or* sentences lie in exhaustive and exclusive *or* inferences. Both exhaustive and exclusive *or* inferences were only occasionally derived for *may/or* sentences as conversational implicatures, but they were derived almost by default for *must/or* sentences as parts of logical interpretations. Let's temporarily not discuss the contrast between *may/or* and *must/or* sentences in the aspect of exclusive *or* inferences (because no definite answer on the nature of exclusive *or* inferences of *must/or* sentences can be given), but only look into the contrast in exhaustive inferences triggered by the two types of sentences. Simons [20] successfully predicts this difference because she assumes that the alternative semantics for disjunction is activated under the scope of deontic modals. Due to this, even if we assume that the sets of alternatives is semantically closed under disjunction, the standard semantics of the deontic possibility modal still opens up the possibility for making *may/or* sentences non-exhaustive. Geurts [12] fails to do so because he claims that disjunction takes scope over deontic modals, and in addition, the list of modal propositions is by default closed under disjunction. Once the list is closed, *may/or* sentences can only be exhaustive. Thus, Simons's [20] account has the highest explanatory power.

To conclude, this study intends to convey three pieces of information: first, the free choice puzzle is better to be approached semantically; second, there is a contrast between *may/or* and *must/or* sentences in deriving exhaustive and exclusive *or* inferences; third, the scope relation between disjunction and deontical modals should be more carefully dealt with.

References

1. Aloni, M.: Free choice, modals, and imperatives. Nat. Lang. Seman. **15**(1), 65–94 (2006)
2. Aloni, M.: Disjunction. In: The Stanford Encyclopedia of Philosophy, Winter 2016 edn (2016). https://plato.stanford.edu/archives/win2016/entries/disjunction/
3. Alonso-Ovalle, L.: Disjunction in alternative semantics. Doctoral dissertation, University of Massachusetts Amherst (2006)
4. Asher, N., Bonevac, D.: Free choice permission is strong permission. Synthese **145**(3), 303–323 (2005)
5. Barker, C.: Free choice permission as resource-sensitive reasoning. Seman. Pragmatics **3**, 10–1 (2010)
6. Bott, L., Noveck, I.A.: Some utterances are underinformative: the onset and time course of scalar inferences. J. Mem. Lang. **51**(3), 437–457 (2004)
7. Chemla, E.: Universal implicatures and free choice effects: experimental data. Seman. Pragmatics **2**, 1–33 (2009)
8. Chemla, E., Bott, L.: Processing inferences at the semantics/pragmatics frontier: disjunctions and free choice. Cognition **130**(3), 380–396 (2014)
9. Chevallier, C., Noveck, I.A., Nazir, T., Bott, L., Lanzetti, V., Sperber, D.: Making disjunctions exclusive. Q. J. Exp. Psychol. **61**(11), 1741–1760 (2008)
10. Fox, D.: Free choice and the theory of scalar implicatures. In: Manuscript, pp. 1–41. MIT (2006)
11. Franke, M.: Quantity implicatures, exhaustive interpretation, and rational conversation. Seman. Pragmatics **4**, 1–1 (2011)
12. Geurts, B.: Entertaining alternatives: disjunctions as modals. Nat. Lang. Seman. **13**(4), 383–410 (2005)
13. Geurts, B.: Quantity Implicatures. Cambridge University Press, Cambridge (2010)
14. Kaufmann, M.: Free choice is a form of dependence. Nat. Lang. Seman. **24**(3), 247–290 (2016)
15. Kratzer, A.: Conditional necessity and possibility. In: Bäuerle, R., Egli, U., von Stechow, A. (eds.) Semantics from Different Points of View, vol. 6, pp. 117–147. Springer, Heidelberg (1979). https://doi.org/10.1007/978-3-642-67458-7_9
16. Kratzer, A.: Modality. In: Von Stechow, A., Wunderlich, D. (ed.) Semantics: An International Handbook of Contemporary Research, pp. 639–50 (1991)
17. Kratzer, A., Shimoyama, J.: Indeterminate pronouns: the view from japanese. In: Otsu, Y. (ed.) The Proceedings of the Third Tokyo Conference on Psycholinguistics. Hituzi Syobo Tokyo, pp. 1–25 (2002)
18. Marty, P.P., Chemla, E.: Scalar implicatures: working memory and a comparison with only. Front. Psychol. **4**, 403 (2013)
19. Sauerland, U.: Scalar implicatures in complex sentences. Linguist. Philos. **27**(3), 367–391 (2004)
20. Simons, M.: Dividing things up: the semantics of or and the modal/or interaction. Linguist. Philos. **13**(3), 271–316 (2005)
21. Starr, W.: Expressing permission. Seman. Linguist. Theory **26**, 325–349 (2016)
22. Van Rooij, R.: Conjunctive interpretations of disjunctions. Seman. Pragmatics **3**, 11–1 (2010)
23. Van Tiel, B.: Universal free choice. In: Proceedings of Sinn und Bedeutung, vol. 16, pp. 627–638 (2012)
24. Zimmermann, T.E.: Free choice disjunction and epistemic possibility. Nat. Lang. Seman. **8**(4), 255–290 (2000)

"First Things First": An Inquisitive Plausibility-Urgency Model

Zhuoye Zhao[✉] and Paul Seip

Institute for Logic, Language and Computation, University of Amsterdam,
Amsterdam, The Netherlands
zhuoye.zhao@student.uva.nl

Abstract. There is a fruitful line of work in incorporating questions into epistemic logic (van Benthem and Minică 2009; Baltag et al. 2016; among others); among others). Based on the viewpoint that communication is a process of raising and resolving issues, inquisitive semantics introduces a uniform notion of meaning for statements and questions, thus can serve as a suitable device for this purpose. For instance, Inquisitive Plausibility Model (Ciardelli and Roelofsen 2014) is able to combine questions with the Epistemic Plausibility Model (IPM) (Baltag and Smets 2006a, b) to capture not only the *belief* and *knowledge* of agents, but also the *issues* they entertain. Building on this, we develop an Inquisitive Plausibility-Urgency Model (IPUM), which not only allows us to model *knowledge*, *belief* and *issues*, but also the *urgency* of the *issues*, hence lead us to towards formalizations of more dynamics of questions.

Keywords: Dynamic epistemic logic · Inquisitive semantics · Plausibility model · Urgency

1 Introduction

A large amount of work has been devoted to the development of logical tools to formally describe and reason about the epistemic change and belief revision through the process of information exchange. A prominent framework that stands out among them is the framework of *dynamic epistemic logic* (DEL) (van Ditmarsch et al. 2007; van Benthem 2007; among others). Such framework is designed to encode a certain set of facts together with what certain agents know or believe about these facts and one another's knowledge and belief. Epistemic situations are then represented by models based on possible world semantics, with the *knowledge* and/or *belief* of an agent captured as a set of propositions (hence, semantically, sets of possible worlds). Epistemic changes and belief revisions of a set of agents can then be captured as *model transformations*. One of the representatives of such models is the Epistemic Plausibility Model (Baltag and Smets 2006a, b), interacting with both *knowledge* and *belief*:

Definition 1. *An Epistemic Plausibility Model $\mathcal{M}$ for a set of agents $\mathcal{A}$ is a tuple $\mathcal{M} = \langle W, \{\leq_a\}_{a \in \mathcal{A}}, \{\sigma_a\}_{a \in \mathcal{A}}, \| \cdot \| \rangle$ where*

© Springer-Verlag GmbH Germany, part of Springer Nature 2019
J. Sikos and E. Pacuit (Eds.): ESSLLI 2018, LNCS 11667, pp. 200–212, 2019.
https://doi.org/10.1007/978-3-662-59620-3_13

- W is a set of possible worlds
- $\leq_a \subseteq W \times W$ is a converse well-founded total preorder between possible worlds encoding a **plausibility map** for each agent
- σ_a is the epistemic map for each agent $a \in \mathcal{A}$. For every $w \in W$, $\sigma_a(w)$ is an equivalence class determined by $\leq_a$: for all $w' \in W$, $w \in \sigma_a(w)$ iff $w \leq_a w'$ or $w' \leq_a w$.
- $\| \cdot \|$ is the valuation function

Some remarks of the definition are in order. For each specific epistemic situation, the plausibility relation $\leq_a$ is set up in a *conditional* manner, namely, fixing any two worlds $w_1, w_2 \in W$, $w_1 \leq_a w_2$ iff the agent a thinks w_2 is at least as *plausible* as w_1, and vice versa. Note that the epistemic map σ_a is defined in terms of the plausibility map $\leq_a$. The reason, conceptual, is that the notion of *plausibility* indicates a certain degree of epistemic indistinguishability between different worlds. For any two worlds $w, w' \in W$, an agent will take them as comparable in terms of plausibility only if they are compatible with her knowledge. Here and henceforth we will keep the notion of epistemic state just for convenience.

The logical language used to talk about epistemic models is a propositional language enriched with several modal operators. For each agent $a \in \mathcal{A}$, there is a knowledge operator K_a, belief operator B_a, strong belief operator Sb_a and conditional belief operator B_a^Q. For an arbitrary proposition φ, the semantics of the modal sentences can be characterized as follows:

- $\mathcal{M}, w \vDash K_a \varphi \Longleftrightarrow \sigma_a(w) \subseteq \| \varphi \|$
- $\mathcal{M}, w \vDash B_a \varphi \Longleftrightarrow Max_{\leq_a} \{ w \in W : w \in \sigma_a(w) \} \subseteq \| \varphi \|$
  (Henceforth, we abbreviate $Max_{\leq_a} \{ w \in W : w \in \sigma_a(w) \}$ as $bel_a(w)$)
- $\mathcal{M}, w \vDash Sb_a \varphi \Longleftrightarrow s >_a t$ for every $s, t \in \sigma_a(w)$ where $s \in \| \varphi \|, t \notin \| \varphi \|$
- $\mathcal{M}, w \vDash B_a^\psi \varphi \Longleftrightarrow bel_a(w) \cap \| Q \| \subseteq \| \varphi \|$

For each agent a, her *knowledge* at w is captured by the epistemic state $\sigma_a(w)$, meaning 'all the possible worlds a knows for sure that's possible at w', whereas her *belief* is captured by $best\sigma_a(w)$, which is determined by $\sigma_a(w)$ and the plausibility relation $\leq_a$, meaning 'the most plausible set of worlds in $\sigma_a(w)$'. The conditional modality B_a^ψ can be read as 'given the information that ψ, the agent a believes...'. Based on this model, the *dynamic* change of the agent's epistemic/doxastic state, i.e. *knowledge update* or *belief revision*, are formalized as model transformers that map the current plausibility model to a new one.

However, there seems to be something missing. As pointed out by Schaffer (2005), '*All knowledge involves a question; To know is to know the answer*'. In modeling the (dynamic) epistemic state of an agent, it is not only important to record the information that is available to the agent, but also the information she would like to obtain, i.e. the *questions*, or the *issues* she entertains. Following the spirit of the 'Socratic epistemology' initiated by Hintikka in the 1970's and later proposed in Hintikka (2007), there is a fruitful line of work in incorporating questions into epistemic/doxastic logic (Olsson and Westlund 2006; Enqvist 2010; among others). With a novel system unifying the semantic characterizations of propositions and questions, Inquisitive Semantics (Ciardelli 2009; Groenendijk and Roelofsen 2009; Ciardelli et al. 2013) managed to enrich the classical models in this aspect (Ciardelli and Roelofsen 2014, 2015). These works elegantly capture some fundamental interactions between the questions an agent entertains and the change of her epistemic state. In this paper, we will provide a modest update of the Inquisitive Plausibility Model (IPM) proposed by Ciardelli and Roelofsen (2014), which enables us to not only incorporate questions into the formalization of epistemic reasoning, but also the *urgency* of each question. In the rest of this section, we present technical details of the background frameworks mentioned above (Sects. 1.1, 1.2 and 1.3), and motivate our model in Sect. 1.4. We will specify our model in Sect. 2, and then show some applications in Sect. 3. We will conclude in Sect. 4.

1.1 Inquisitive Semantics (InqB) and Inquisitive Epistemic Logic (IEL)

Inquisitive semantics (InqB) starts from the observation that the primary function of natural language is to exchange information, thus motivates a notion of meaning that captures not only informative, but also inquisitive content. To achieve this, inquisitive semantics generalizes the meaning of a sentence as the *issue* it raises. An *issue* is characterized as the set of *information states* that *resolve* it. In possible world semantics, an *information state* is formalized as a set of possible worlds (the same as a *proposition*), thus an *issue* can be represented as a *set of sets of possible worlds*. Moreover, if a certain proposition p resolves an issue, any stronger proposition q ($q \subseteq p$) also resolves the issue. Therefore, we require the set of information states of which an *issue* consists of to be non-empty and *downward-closed*. In possible world semantics, a proposition p is *true* in a world w just in case $w \in p$. In parallel, an issue P is *supported/resolved* by a proposition p just in case $p \in P$. Here and henceforth, we will denote both the *truth* relation and the *support* relation as '$\vDash$'.

The maximal elements of an issue P are referred to as its *alternatives*, written as $alt(P)$. That is, $alt(P) := \{A \in P \mid \forall B \in P : A \not\subseteq B\}$. A sentence is *inquisitive* if it has more than one alternative, and is *non-inquisitive* if it has only one alternative. The *information content* of an issue P, denoted by $|P|$ or $info(P)$, is defined as the union $\bigcup P$ of all elements in P.

The language of InqB is very much like propositional logic, with atomic formula $p, q...$, negation $\neg$, boolean connectives $\vee, \wedge, \rightarrow$ of similar accounts, except for two additional projection operators, ! and ?, which are referred to as *non-inquisitive* and *non-informative operators*, respectively. The non-inquisitive operator ! maps an issue P to the power set of its informative content, i.e. $!P = \mathcal{P}(|P|)$, while the non-informative operator ? maps P to the disjunction of itself and its negation, i.e. $?P = P \cup S \backslash |P|$.

Based on InqB, Inquisitive Epistemic Logic (IEL) extends the classical epistemic logic with additional inquisitive notions. IEL introduces two basic epistemic notions, the *epistemic state* $\sigma_a(w)$ capturing the knowledge of an agent a at a world w (same as the classical notion); and the *inquisitive state* $\Sigma_a(w)$ that captures the issue the agent a *entertains*. For any information state $s \in \Sigma_a(w)$, s is a resolution to the issue that concerns a. Moreover, the epistemic state of an agent is always equivalent to the informative content of its inquisitive state, that is, $\sigma_a(w) = \bigcup \Sigma_a(w)$. With these basic notions, we can then introduce two basic epistemic modalities that we will operate on.

Definition 2. *The knowledge modality* K_a

$$w \vDash K_a \varphi \Longleftrightarrow \sigma_a(w) \in \varphi$$

That is, an agent *knows* a sentence φ if and only if the agent's epistemic state resolves the issue raised by the sentence. Similarly, we can define a modality E_a that pictures the issues that the agent *entertains*.

Definition 3. *The Entertain modality* E_a

$$w \vDash E_a \varphi \Longleftrightarrow \forall s \in \Sigma_a(w), s \in \varphi$$

1.2 Inquisitive Plausibility Model (IPM)

Based on classical plausibility model, InqB and observations made in Olsson and Westlund (2006), Enqvist (2010), Ciardelli and Roelofsen (2014) proposed a semantic framework known as inquisitive plausibility model to capture not only the belief and knowledge of an agent, but also the issues an agent entertains, which can be viewed as her "long-term epistemic goals". Further, it can be used to model the *research agenda* of an agent, which is captured as the issues the agent entertains conditioning on her belief. The formalizations are as follows.

Definition 4. *An inquisitive plausibility model for a set of agents $\mathcal{A}$ is a tuple* $\langle W, \| \cdot \|, \{\sigma_a\}_{a \in \mathcal{A}}, \{\leq_a\}_{a \in \mathcal{A}}, \{\Sigma_a\}_{a \in \mathcal{A}} \rangle$ *that consists of:*

- *a set W of possible worlds.*
- *a valuation function $\| \cdot \|$.*
- *an epistemic map σ_a for each agent.*
- *a plausibility map $\leq_a$ for each agent.*
- *an inquisitive map Σ_a for each agent $a \in \mathcal{A}$: for every $s \in S$, $\Sigma_a(s)$ is an issue over $\sigma_a(s)$.*

The language and semantics of a corresponding inquisitive belief logic is then naturally adapted from IEL and classical plausibility model. In addition to the knowledge and belief modalities defined above, we further enrich the language with modalities of *entertaining* E_a, *conditional entertaining* E_a^ψ and *entertaining-over-belief* E_a^B, whose semantics are given as below. Here we denote information states by α and issues by $\mu :=?\{\alpha_1, \ldots, \alpha_n\}$. Also, we will always assume that μ is in the minimal form, i.e. $\alpha_1, \ldots, \alpha_n$ are non-redundant alternatives, and μ is the downward closure of them.

Resolution
  $M, w \vDash ?\{\alpha_1, \ldots, \alpha_n\} \Longleftrightarrow$ for some α_i, $M, w \vDash \alpha_i$ for every $w \in s$.

Truth Conditions

- $\mathcal{M}, w \vDash E_a\mu \iff \forall t \in \Sigma_a(w), t \in \mu$
- $\mathcal{M}, w \vDash E_a^\psi\mu \iff \forall t \subseteq \| \psi \|, t \in \Sigma_a(w) \Rightarrow t \in \mu$
- $\mathcal{M}, w \vDash E_a^B\mu \iff \forall t \subseteq bel_a(w), t \in \Sigma_a(w) \Rightarrow t \in \mu$

Note that the entertain-over-belief modality E_a^B is used to address issues the agent is entertaining over her beliefs. We will refer to the set of such issues as the agent's *research agenda*.

1.3   Inquisitive Contractions

Here we introduce an application of IPM. Classical belief revisions are captured as model transformations between plausibility models. Using IPM, we are not only able to preserve the classical operations, but also successfully model changes of an agent's research agenda. One of the typical phenomena as such is known as *inquisitive contractions* (Olsson and Westlund 2006). To keep it straightforward, we will elaborate with the following single-agent scenario, and show how IPM can be used to model this dynamic process.

Scenario: Alice believes there is a seminar this afternoon. She is wondering if Bill is coming. Suddenly Charlie pops up and says to her that there won't be a seminar today. Alice doesn't totally buy it, yet she also starts wondering if there will be a seminar this afternoon.

Analysis: The inquisitive state of Alice was the issue $?S \wedge ?Cb$. However, since she believed that there is a seminar (S), she was actually considering whether Bill is coming at first, as shown in Fig. 1(a). After being informed by Charlie, her belief is weakened – she thinks it is quite possible that there won't be a seminar $([\downarrow S])$, hence her belief state fully covers her epistemic state, and she goes back to entertain the issue $?S \wedge ?Cb$, as in Fig. 1(b).

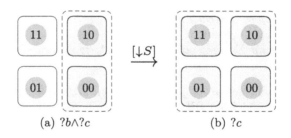

(a) $?b \wedge ?c$ (b) $?c$

Fig. 1. Contraction

Therefore, IPM can be used to model inquisitive contractions, and in general, the dynamics of belief changes interacting with the agents' research agenda. However, there is more to be done to increase the descriptive power. First let us consider two motivating scenarios.

1.4   Motivating Scenarios: First Things First!

We have seen from the previous sections the basic setup of the inquisitive plausibility model (IPM) and how it captures a fundamental (and common) interaction between questions and belief change, namely the inquisitive contraction. In this section, we will motivate a modest update to IPM with two scenarios requiring a more fine-grained structure over the *research agenda*, i.e. an *urgency* order. We will see in later sections that it can be (quite easily) implemented and potentially extend the coverage of current models over real life scenarios.

Scenario 1: Alice asked Bill and Charlie to have dinner in a restaurant. Alice doesn't know whether they will come, and she wants to know. Moreover, Alice knows Bill likes Pasta, and Charlie likes rice. She knows that both prices of rice and pasta are 7 or 8 euros, but she is not sure which is which, and she wants to know. Then Bill called, telling her that he is coming, but Charlie is not. Now Alice feels more *urgent* to know the price of pasta.

Scenario 2: Alice believes that there is a seminar this afternoon, and she is wondering whether Bill is coming. Now Charlie pops up again and says to her that there won't be a seminar today. Alice doesn't totally buy it, yet now she wants to know whether there is a seminar *first*.

Here and henceforth we will refer to Scenario 1 as the "dinner scenario" and Scenario 2 as "seminar scenario". In the dinner scenario, the belief update of Alice (namely, Bob is coming but Charlie is not) resolves the issue of whether Bill or Charilie is coming, and further gives rise to different *urgencies* between the other two questions: the price of pasta is now more relevant than the price of rice as Bill likes pasta. In the seminar scenario, the contraction of Alice's believe brings back the question of whether there is a seminar, which is then more *urgent* than the question of whether Bill is coming. As is shown in both cases, a formalization of the notion of *urgency* of questions is in need to capture the change of the epistemic/doxastic state of Alice and her further inquiry of information. IPM can capture the raising and resolving of issues followed by knowledge/belief change, but fails to provide the more fine-grained urgency relation between the different issues in the research agenda. In the next section, we will introduce the *Inquisitive Plausibility-Urgency Model*, which is a modest extension to IPM, and will thus provide us with more dynamics for belief revision and questions.

2   Inquisitive Plausibility-Urgency Model (IPUM)

2.1   The *Urgency* Relation

As mentioned in the final part of last section, our goal here is to (qualitatively) model the degree of *urgency* of different questions. How can we achieve it? Recall that in Epistemic Plausibility Model, the meaning of a proposition is characterized as the set of possible worlds on which the proposition is true, and we capture the agents' *beliefs* by defining a plausibility relation between possible worlds, which is based on an agent's attitudes toward different worlds. In inquisitive semantics, the meaning of a question is modeled as the set of information states that resolve it, therefore in accordance, we should be able to model an agent's attitude towards a question as a collective manifestation of her attitude towards its resolutions. Moreover, in practice, one is more *urgent* to know the answer of a question means he/she will feel more satisfied/relieved after hearing it. Therefore, just like the plausibility relation, we can define an *urgency* relation $\preccurlyeq_a$ between information states in a *conditional* manner, as follows. Fixing any two information states s, t:

$$s \preccurlyeq_a t \Longleftrightarrow a \text{ is at least as urgent/satisfied/relieved to know } t \text{ as she is for } s$$

By adding this urgency relation to IPM, we will get what we want – the *Inquisitive Plausibility-Urgency Model*.

2.2   The Model

Definition 5. *An inquisitive plausibility-urgency model $\mathcal{M}$ for a set of agents $\mathcal{A}$ is a tuple*

$$\mathcal{M} = (W, \{\sigma_a\}_{a \in \mathcal{A}}, \{\Sigma_a\}_{a \in \mathcal{A}}, \{\leq_a\}_{a \in \mathcal{A}}, \{\preccurlyeq_a\}_{a \in \mathcal{A}}, \| \cdot \|)$$

where

- *W is a set of possible worlds.*
- *an epistemic map σ_a for each agent.*
- *an inquisitive map Σ_a for each agent.*
- *a plausibility map $\leq_a \subseteq W \times W$ for each agent.*
- *an urgency map $\preccurlyeq_a \subseteq \mathcal{P}(W) \times \mathcal{P}(W)$ for each agent a at w, which is a converse-well-founded total preorder on information states. Also, for each $s, t \in \mathcal{P}(W)$ s.t. $s \preccurlyeq_a t$, if $t' \subseteq t$, then $s \preccurlyeq_a t'$.*
- *a valuation function $\| \cdot \|$.*

Similar to Epistemic Plausibility Model as in Definition 1, we introduce modalities of *urgent-entertaining* (UE_a), *strong urgent-entertaining* (Sue_a) and *conditional urgent-entertaining* UE_a^ψ. Given some issue $\mu := ?\{\alpha_1, ..., \alpha_n\}$, their semantic characterizations are as follows:

- $\mathcal{M}, w \vDash UE_a \mu \Longleftrightarrow$ for any $\alpha \in Max_{\preccurlyeq_a}\{t \in \mathcal{P}(W) : t \in \Sigma_a(w)\}, \alpha \in \mu$
- $\mathcal{M}, w \vDash Sue_a \mu \Longleftrightarrow$ for any $s, t \in \mathcal{P}(W)$, if $s \notin \mu$ and $t \in \mu$, then $s \preccurlyeq_a t$.
- $\mathcal{M}, w \vDash UE_a^\psi \mu \Longleftrightarrow$ for any $t \in Max_{\preccurlyeq_a}\{t \in \mathcal{P}(W) : t \in \Sigma_a(w)\}$ and $t \subseteq \| \psi \|$, $t \in \mu$
- $\mathcal{M}, w \vDash UE_a^B \mu \Longleftrightarrow$ for any $t \in Max_{\preccurlyeq_a}\{t \in \mathcal{P}(W) : t \in \Sigma_a(w)\}$ and $t \subseteq bel_a(w), t \in \mu$

Also similar to E_a^B, UE_a^B is the urgent-entertaining-over-belief modality, which is used to address issues the agent is urgently entertaining over the current belief state.

Some remarks need to be made here. First, the inquisitive map Σ_a takes a world w and returns the *total issue* the agent a entertains in the world w. That is, any information state that resolves $\Sigma_a(w)$ answers all the questions a has at w. Second, when constructing a model regarding specific situations, we take complete resolutions to an issue that is **urgently** entertained as equivalently satisfying, i.e. given $\mu \; =?\{\alpha_1, ..., \alpha_n\} \in \Sigma_a(w)$ and $\mathcal{M}, w \vDash UE_a\mu$, for any $s, t \in \mu$, there is $s \preccurlyeq_a t$ and $t \preccurlyeq_a s$, or rather, $s \approx_a t$, where $\approx_a$ is the equivalence relation in terms of the urgency order.[1] With this condition at hand, as the case in Epistemic Plausibility Model that $\leq_a$ can fully describe the Epistemic map σ_a, the urgency relation $\preccurlyeq_a$ also determines the inquisitive state Σ_a, in the sense that the degree of urgency indicates certain inquiries. Formally, it can be achieved by simply intersecting all the downward closures of the equivalent sets partitioned on $\mathcal{P}(W)$ by $\approx_a$ (excluding the least urgent one containing W). Last but not least, potential changes of research agendas, or in a sense the inquiry strategies are pre-encoded in the model. In particular, differences in urgency may be revealed at different level of information, which result in a more fine-grained reaction of an agent towards different information pieces. This feature can lead us to a solution to the dinner scenario, which requires different reactions of Alice given different information (whether Charlie or Bill is coming). We will show how IPUM provides us with a relatively elegant formalization of both scenarios mentioned in Sect. 1.4 in the next section. Before that, we will first introduce a bonus effect coming with IPUM.

2.3 Constructing the Agenda

In this section we introduce a technical advantage resulting from the addition of urgency relation. In IPM and other models combining questions with Epistemic logic, the issue an agent is entertaining is captured as a total question, or a 'long-term epistemic goal', whereas the total question may consist of many small questions. As stated in Baltag et al. (2016), an agent may have a set of *fundamental issues* about all world-properties that she considers relevant and wishes to resolve, and the conjunction of all fundamental questions amounts to the *total issue* that captures the agent's ultimate epistemic goal. As we can see, the inquisitive map Σ_a has already captured the notion of *total issue*. With the additional urgency relation $\preccurlyeq_a$ and under the condition that all the fundamental issues are strictly ordered, we can further formalize the notion of *fundamental issues*, which is defined as follows. Here we use the term *cluster* to refer to a equivalent set w.r.t. the urgency relation (of an agent), i.e. a cluster $C \subseteq W$ satisfies the condition that for each $w, w' \in C$, $w \preccurlyeq_a w'$ and $w' \preccurlyeq_a w$.

Definition 6. $F_C \in \mathcal{P}(S)$ *is a fundamental issue if there is a cluster C for which $F_C = \downarrow C$. Particularly, if there is $u \in C$ such that $u \notin \Sigma$, then $\Sigma \subsetneq F_C$.*

---

[1] Note that this condition only applies to urgently entertained issues. The reason is, for any issue μ that is not urgently entertained, an information state α that resolves both μ and an urgently entertained issue μ' would be preferred than any other α' that only resolves μ.

The additional notion in Definition 6 means that if there are multiple fundamental questions, resolutions to any fundamental question may not necessarily resolve the total issue. In IPUM, the urgency relation manifests itself between clusters, hence there should be a correspondence between clusters and fundamental questions. Under the condition that the fundamental issues are strictly ordered, this correspondence is one-to-one. Therefore, we can retrieve the fundamental issues simply by taking downward closure of clusters.[2] Now let's illustrate how IPUM retrieves the fundamental issues through a party scenario.

Scenario: Alice was invited to a party, and she knew that Bill and Charlie were also invited. Alice wonders whether Bill will come $(?b)$, and whether Charlie will come $(?c)$, but she considers Bill a better friend than Charlie. As a consequence Alice is more interested to know whether Bill is coming.

Analysis: Here Alice has two fundamental issues $?b$ (whether Bill is coming) and $?c$ (whether Charlie is coming). She has different *urgencies* of resolving them, with the resolutions to $?b$ more urgent than the resolutions to $?c$. The IPUM is sketched in Fig. 2. Notice that Fig. 2(a)–(c) pictures the maximal elements in clusters (here we assume resolutions to the total issue is in the equivalent set corresponding to $?b$, for a better illustration). By taking the downward closure, we get the list of fundamental issues $\mathcal{F} = \{?c, ?b\}$, and the total issue, as pictured in Fig. 2(d), is the intersection of all the fundamental issues.

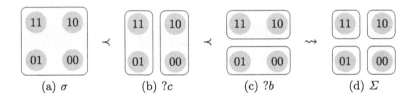

Fig. 2. Constructing the agenda

3   Towards More Dynamics of Questions

In this section, we will go through some applications of IPUM, by resolving the motivation scenarios introduced in Sect. 1.4. By resolving the dinner issue, we demonstrate that the urgency order among different questions can be revealed

---

[2] Retrieving the fundamental issues are actually the major motivation for the original paper. However, we noticed, as was also kindly pointed out by Dr. Floris Roelofsen, this can only be done under the condition that the clusters are strictly ordered. This result is really unsatisfying. However, this model is successful in capturing reactions towards partial resolutions, and by altering our motivation, we get to the conditional definition of urgency relation, which is also more intuitive.

on partial resolutions to the total issue, therefore capture the reaction of an agent toward certain pieces of information. The solution to the seminar scenario will present a more fine-grained formalization of inquisitive contraction. Starting from these applications, it is straightforward to extend to more dynamics of questions interacting with knowledge/belief change using IPUM.

Scenario 1: Alice asked Bill and Charlie to have dinner in a restaurant. Alice doesn't know whether they will come, and she wants to know. Moreover, Alice knows Bill likes Pasta, and Charlie likes rice. She knows that both prices of rice and pasta are 7 or 8 euros, but she is not sure which is which, and she wants to know. Then Bill called, telling her that he is coming, but Charlie is not. Now Alice feels more *urgent* to know the price of pasta.

Analysis 1: Initially, Alice was wondering whether Bill and Charlie was coming, as well as the price of rice and pasta. She had neither belief nor urgency on either of the two issues, therefore her research agenda is the total issue as pictured in Fig. 3. Note that here we illustrate the four fundamental issues *separately*: Fig. 3(a) represents the issue whether Bill is coming, Fig. 3(b) whether Charlie is coming, Fig. 3(c) the price of pasta and Fig. 3(d) the price of rice. Only the alternatives of these issues are illustrated here. The grey circles with $0/1$ (or $0'/1'$) pairs should not be read as worlds here, rather, they are (maximal) information states that resolve both the issues illustrated in Fig. 3(a) and (b) (or Fig. 3(c) and (d)). The total issue, then, should be the downward closure of the intersection of all the four fundamental issues, with 16 alternatives (e.g. $11 \times 1'1'$ represents the alternative supporting the fact that both Charlie and Bill are coming and both pasta and rice are 7 euros, etc.). Assuming the alternatives within each fundamental issue are equivalently urgent, the urgency relation between different fundamental issues given in Fig. 3 (in this case, all are equivalent to a) indicates that resolving any one of the fundamental issue (but not more) is equivalently urgent, capturing the initial status of Alice's epistemic state.

Since Alice knows that Bill likes Pasta and Charlie likes rice, the urgency relation between resolutions to the price issues should vary given information that resolves the issue of whether Bill and Charlie are coming. For instance, under the condition that Bill is coming but Charlie is not ($Cb \land \neg Cc$), Alice should be more eager to know the price of pasta. Therefore, we can construct a strict urgency order between resolutions to the price issue among information states that are already resolutions to the coming issue, as shown in Fig. 3. Note that the segment by $0'/1'$ pairs in Fig. 4 is restricted on the set of worlds where Bill is coming and Charlie is not (i.e. the 10 worlds in Fig. 3). After Bill called, Alice came to believe that Bill was coming but not Charlie, therefore Fig. 4(a) pictures just the research agenda of Alice at that moment, where the price of pasta is strictly more urgent than the price of rice. Based on this model, the following propositions are true:

- $\mathcal{M}, w \vDash E_a(?C_b \land ?C_c \land (P_7 \lor P_8) \land (R_7 \lor R_8))$
- $\mathcal{M}, w \vDash UE_a^{C_b \land \neg C_c}(P_7 \lor P_8)$
- $\mathcal{M}, w \vDash Sue_a^{C_b \land \neg C_c}(P_7 \lor P_8)$

That is, (i) Alice has an inquisitive state consisting of both coming issues and price issues; (ii) Conditioning on the information that Bill is coming but Charlie is not, Alice will urgently entertain the price of pasta, and (iii) in fact, she will "strongly urgently" entertain this issue. Similarly, we can imagine the situation where Bob is not coming and Charlie is coming, knowing the price of rice should be more urgent for Alice than the price of pasta, and in the situation where both or neither of them are coming, the price issues are equivalently urgent. Crucially, the model captures the fact that none of the four fundamental issues in Fig. 3 is *globally* more urgent than any other; it is certain resolutions of the coming issue that bring out (*locally*) strict urgency orders between the price issue. Hence we capture the interaction between belief update and research agenda described in the dinner scenario.

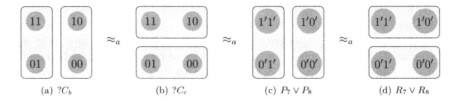

Fig. 3. Dinner scenario - before

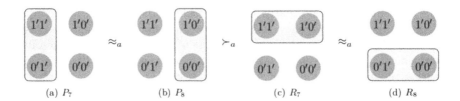

Fig. 4. Dinner scenario - after

Scenario 2: Alice believes that there is a seminar this afternoon, and she is wondering whether Bill is coming. Now Charlie pops up again and says to her that there won't be a seminar today. Alice doesn't totally buy it, yet now she wants to know whether there is a seminar *first*.

Analysis 2: Alice's epistemic goal is to know whether there is a seminar today and whether Bill is coming. Moreover, Alice thinks the former as the more urgent issue. Therefore her inquisitive state along with the urgency order can be shown as in Fig. 5(a). However, initially Alice believed that there is a seminar, which makes the latter the only issue she is entertaining (see also Fig. 1). After withdrawing this belief, the urgency order reveals itself. Hence we have the following results:

- $\mathcal{M}, w \vDash E_a^B(?Cb)$
- $\mathcal{M}, w \vDash [\downarrow S]UE_a(?S)$

That is, based on her belief, Alice is entertaining whether Bill is coming, but when the belief is retracted, she begins to urgently entertain whether there is a seminar.

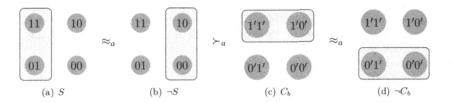

(a) S (b) $\neg S$ (c) C_b (d) $\neg C_b$

Fig. 5. Seminar scenario

The above analyses show that with the addition of the urgency relation defined in IPUM, we can capture more interactions between questions and belief update where the change of epistemic/doxastic state gives rise to an urgency order on the agent's research agenda.

4   Conclusion

In this paper, we proposed an extension of inquisitive plausibility model named inquisitive plausibility-urgency model by introducing an *urgency* relation between information states. The relation is defined in a conditional manner analogous to the plausibility relation. Using this model, we can capture more dynamics of questions in belief change, such as the reaction of an agent towards (partial) resolutions to the issue she's entertaining, as well as a more fine-grained notion of inquisitive contraction.

The description of IPUM in the paper is rather sketchy, and we hope to further specify the details in a full paper. Meanwhile, based on this basic model, some immediate future direction should be concerned. For one thing, a sound and complete axiomatization (logic) is open for investigation. For another, here we restrict our model in a single-agent setting; in order to extend it to a multi-agent setting, we need to consider additional complications such as the attitudes of an agent towards the issues raised or entertained by other agents, which will hopefully lead us to a more complete picture of insecure communication. We hope this paper can serve as a modest spur that induces revelations of inquisitive potential, as well as its influence on dynamic epistemic/doxastic logic.

Acknowledgements. The authors would like to thank Dr. Alexandru Baltag for the course *Dynamic Epistemic Logic* at the University of Amsterdam. We are also grateful to Dr. Floris Roelofsen, the Inquisitive Semantics group at the ILLC, and the anonymous reviewers for their valuable feedbacks. Last but not least, many thanks to ESSLLI 2018 Grant Committee and EACL for the student grants.

References

Baltag, A., Boddy, R., Smets, S.: Group knowledge in interrogative epistemology. In: van Ditmarsch, H., Sandu, G. (eds.) Jaakko Hintikka on Knowledge and Game-Theoretical Semantics. Springer, Cham (2016). https://doi.org/10.1007/978-3-319-62864-6_5

Baltag, A., Smets, S.: Conditional doxastic models: a qualitative approach to dynamic belief revision. Electron. Notes Theor. Comput. Sci. **165**, 5–21 (2006a)

Baltag, A., Smets, S.: The logic of conditional doxastic actions: a theory of dynamic multi-agent belief revision. In: Proceedings of ESSLLI Workshop on Rationality and Knowledge, pp. 13–30 (2006b)

van Benthem, J.: Dynamic logic for belief revision. J. Appl. Non-Class. Log. **17**(2), 129–155 (2007)

van Benthem, J., Minică, Ş.: Toward a dynamic logic of questions. In: He, X., Horty, J., Pacuit, E. (eds.) Logic, Rationality, and Interaction, pp. 27–41. Springer, Heidelberg (2009). https://doi.org/10.1007/978-3-642-04893-7_3

Ciardelli, I.: Inquisitive semantics and intermediate logics. Master thesis, University of Amsterdam (2009)

Ciardelli, I., Groenendijk, J., Roelofsen, F.: Inquisitive semantics: a new notion of meaning. Lang. Linguist. Compass **7**(9), 459–476 (2013)

Ciardelli, I., Floris R.: Issues in epistemic change. Talk at European Epistemology Network Meeting, Autonomous University of Madrid, July 2014 (2014)

Ciardelli, I.A., Roelofsen, F.: Inquisitive dynamic epistemic logic. Synthese **192**(6), 1643–1687 (2015)

van Ditmarsch, H., van Der Hoek, W., Kooi, B.: Dynamic Epistemic Logic, vol. 337. Springer, Dordrecht (2007). https://doi.org/10.1007/978-1-4020-5839-4

Enqvist, S.: Contraction in interrogative belief revision. Erkenntnis **72**(3), 315–335 (2010)

Groenendijk, J., Roelofsen, F.: Inquisitive semantics and pragmatics. In: Larrazabal, J.M., Zubeldia, L. (eds.) Meaning, Content, and Argument: Proceedings of the ILCLI International Workshop on Semantics, Pragmatics, and Rhetoric (2009). www.illc.uva.nl/inquisitive-semantics

Hintikka, J.: Socratic Epistemology: Explorations of Knowledge-Seeking by Questioning. Cambridge University Press, Cambridge (2007)

Olsson, E.J., Westlund, D.: On the role of the research agenda in epistemic change. Erkenntnis **65**(2), 165–183 (2006)

Schaffer, J.: Contrastive knowledge. Oxf. Stud. Epistemol. **1**, 235–271 (2005)

Interpreting Intensifiers for Relative Adjectives: Comparing Models and Theories

Zhuoye Zhao[✉]

Institute for Logic, Language and Computation,
University of Amsterdam, Amsterdam, The Netherlands
zhuoye.zhao@student.uva.nl

Abstract. Adjectives such as *tall* or *late* which can enter comparative constructions or be modified by intensifiers such as *very* are called *gradable*. They have received considerable attention in formal semantics and, more recently, in Bayesian pragmatics. While comparative constructions are well understood, less is known about the contribution of intensifiers. In this paper, we compare several concrete models for the meaning of *very tall* with data from a previous study, in the hope of shedding light on the semantic contributions of intensifiers.

Keywords: Degree semantics · Gradable adjectives ·
Bayesian pragmatics · Degree modifiers

1 Introduction

1.1 Degree Semantics

Gradable adjectives are adjectives, such as '*tall*' or '*late*', that can appear in comparative ('*taller*', '*later*') and superlative ('*tallest*', '*latest*') constructions and can be modified by intensifiers (e.g. '*very tall*'). Degree Semantics (Kennedy and McNally 2005; Kennedy 2007; among others) proposes that the meaning of gradable adjectives can be characterized as a function that maps individuals to *degrees* on a *scale*. For example, the meaning of '*tall*' can be denoted as $[\![tall]\!] = \lambda x.\mathbf{height}(x)$, which gives a map from individuals to their degree of *tallness*. When used in positive forms as in *Ronald is tall*, which is obtained by combining *tall* with a null morpheme *pos*, it means the *degree* of *tallness* (i.e. height) exceeds a context-dependent threshold $\theta(C)$.

(1)　$[\![pos\ tall]\!] = \lambda x.\mathbf{height}(x) \geq \theta(C)$

where the threshold $\theta(C)$ is determined contextually in the sense that it has different values with respect to different *comparison classes* (encoded by the argument C). For instance, *a tall tree* and *a tall man* are judged by different standards. The former is based on a comparison class consisting of the heights of trees,

© Springer-Verlag GmbH Germany, part of Springer Nature 2019
J. Sikos and E. Pacuit (Eds.): ESSLLI 2018, LNCS 11667, pp. 213–224, 2019.
https://doi.org/10.1007/978-3-662-59620-3_14

whereas the latter the heights of adult males. The Bayesian models we will introduce shortly show ways in which the threshold is selected according to C.

Different adjectives differ in the type of measure functions they denote and the associated scale structures. Kennedy and McNally (2005) distinguished between *absolute adjectives* like '*late*' or '*full*', which map individuals onto *closed* degree scales with lower- or upper-bounds, and *relative adjectives* like '*tall*' with *open* scales. Scale structure affects $\theta(C)$: with a closed scale, absolute adjectives tend to pick the scale endpoint as their threshold, which is known and does not depend much on the comparison class; whereas open scales do not provide such a salient threshold, so their $\theta(C)$ is more context-dependent and less certain. As a consequence, relative adjectives tend to be *vague*. The vagueness manifests itself at so-called *borderline cases* Williamson (2002). For example, it is hard to decide whether a 5 ft 10 in. tall man is tall. By contrast, adjectives like '*late*' or '*full*' show no such vagueness – someone is *late* as long as they show up after the scheduled time[1].

Having introduced a comparison threshold $\theta(C)$ into the semantics of gradable adjectives, the meaning of degree modifiers such as *very* follows naturally – they shift the threshold $\theta(C)$ to a higher value. Klein (1980) proposed a formal semantic account for degree modifiers that captures the threshold shifting, which can be easily adapted into the notions of degree semantics: the basic idea is that a sentence like *Ronald is very tall* is true if Ronald is tall compared to the set of tall people. We can formally define it as follows:

(2) $[\![\text{very } pos \text{ tall}]\!] = \lambda x.\mathbf{height}(x) \geq \theta(C')$,
$$\text{where } C' = \{x \mid [\![pos \text{ tall}]\!](x) = 1\}$$

One of the goals of this paper is to test the prediction of this account using probabilistic pragmatic models (see Sect. 1.2), and compare it with another purely pragmatic account proposed by Bennett and Goodman (2018) (also see Sect. 1.2).

---

[1] In practice we might use these expressions *loosely*, e.g. if Ronald arrives just a few seconds later, it may not really count as *late*. But here we should distinguish between *imprecision* and *vagueness* (see also Kennedy and McNally 2005). We can read an adjective as *absolute* as long as we are able to set a fixed standard of comparison *if forced to*, and even without reference to the context (e.g. '*full*', '*empty*').

However, this characterization of the absolute-relative distinction is based on the assumption that a predicate is vague as long as it has borderline cases. But it has been argued that this conception might not be sufficient (e.g. Smith 2008 Sect. 3.2). An alternative characterization, as noted by Burnett (2014), Egré and Bonnay (2010), would be the *asymmetric tolerance* displayed by absolute adjectives (but not relative adjectives): If Ronald is only a few seconds late, it might still count as *not late* in certain contexts; in contrast, if he is a few seconds earlier, then he is definitely not late. Such asymmetry is also visible in the experimental data we refer to later Leffel et al. (2018). Since this paper targets mainly on relative adjectives, we will not further this discussion (with an exception in Sect. 2.5), but it would be ideal for any quantitative models of gradable adjectives to capture such pattern. Many thanks to the anonymous reviewer for bringing this point to my attention.

1.2   Probabilistic Pragmatic Models for Gradable Adjectives

In order to fully capture the meaning of gradable adjectives, a mechanism to determine or infer the context-dependent threshold $\theta(C)$ is necessary. Fortunately, as mentioned above, probabilistic pragmatic models (in the sense of Franke and Jäger 2016) have been successful in making quantitative predictions for $\theta(C)$, and in further explaining the linguistic phenomena we are interested in. The basic assumption of such models, as a combination of Gricean reasoning and Bayesian modeling, is that language users are goal-oriented Bayesian agents that are involved in social interactions where speaker and listener communicate and recursively reason about each other's goals and inferences. Each utterance u is attributed probability $P(u|w)$ to be chosen by a speaker with knowledge-state w under the assumption that the speaker is trying to maximize some notion of utility (the definition of which varies in different models). Then the Listener applies Bayes' rule to obtain a probability distribution on possible states of the world given what the speaker said.

Here we present two probabilistic pragmatic models for gradable adjectives: the Rational Speech-Act Model (RSA) proposed by Lassiter and Goodman (2014), and the Speaker-Oriented Model (SOM) by Qing and Franke (2014). The RSA model is a *listener-oriented* model, which predicts the threshold $\theta(C)$ as an inference of a (pragmatic) listener, whereas the SOM model, as indicated by the name, derives $\theta(C)$ at the speaker's level based on his/her prior knowledge about the world. Therefore, these models are Bayesian in a strong sense (Franke and Jäger 2016) as they not only represent speakers' subjective uncertainty in terms of probability distributions, they also describe the listeners' interpretation in a form of abductive reasoning with Bayes' rule. We will now present the main features of these two models. In the following we focus on the relative adjective *tall* and the utterance '*Ronald is tall*', marked as u_1. We take u_0 as the empty utterance.

Rational Speech-Act Model (RSA)

The essential idea behind (Gricean inspired) Bayesian models of pragmatics is that the listener uses Bayes' rule to recover a speaker's knowledge state w in a context C given the speaker's utterance u.

(3) $P(w|u, C) \propto P(u|w, C) \times P(w)$

In our case, we are only interested in Ronald's height h_0, so w can be reduced to the speaker's knowledge of this height. We assume that the listener has a prior knowledge of the distribution of possible heights, $\phi(h)$. The only missing part now is a model that derives the speaker's choice of utterance, characterized as a probability measure $P(u|w, C)$.

The idea behind RSA is that the listener models a speaker who tries to minimize the *cost* of his/her utterance, while maximizing the *informativity* for a virtual "literal listener" L_0, who simply updates the prior by conditioning on u being literally true. This is where the semantics come into play: Lassiter and Goodman (2014) assume the standard Degree semantics, so the truth of u depends on a threshold θ. In their model, the listener has no prior knowledge of θ, but assumes that the speaker has exact knowledge of it. θ is thus a free parameter that the listener must infer together with h_0, i.e. Ronald's actual height.

Concretely, informativity is measured as the negative surprisal value of L_0's belief about h_0 after hearing u, whereas the cost is described as a function over the domain of possible utterances u. Then we can define the following *utility* function of possible utterances:

(4) $U_{\text{rsa}}(u, \theta, h_0) = \log(\phi(h_0|u, \theta)) - \text{Cost}(u)$

The speaker tries to maximize the utility, but is assumed to do it in a sub-optimal fashion. The reason behind this additional requirement is that real-life speakers are assumed to be not perfectly rational, thus may make mistakes and deviate from the optimal choice. Using the standard soft-max function with a parameter $\lambda \in [0, +\infty)$ measuring the degree of rationality, we can then derive the probability function of the speaker's choice of utterance:

(5) $\sigma(u|\theta, h_0) \propto exp(\lambda \cdot U_{\text{rsa}}(u, \theta, h_0))$

The listener then infers a joint distribution for θ and h by applying Bayes' rules. Here we will only be interested in the posterior distribution of θ, which is given by the formula in (6).

(6)   With $c_{\text{rsa}} = \text{Cost}(u)$ and Pr the (uninformative) prior on θ:

$$\rho(\theta|u) \propto \int_{-\infty}^{\infty} \phi(h) \cdot Pr(\theta) \cdot \sigma(u|h, \theta)dh = \frac{Pr(\theta) \cdot \int_{\theta}^{\infty} \phi(h)dh}{1 + e^{\lambda c_{\text{rsa}}} \cdot (\int_{\theta}^{\infty} \phi(h)dh)^{\lambda}}$$

Speaker-Oriented Model (SOM)

The SOM differs from the RSA model in that instead of the literal listener L_0, it assumes a listener L sharing the prior knowledge $\phi(h)$ with the speaker. Moreover, instead of a fixed value of the threshold $\theta(C)$, it provides a mechanism to derive the probability of the speaker using a specific θ, hence gives a generalization over possible contexts.

Concretely, keeping the assumption that the speaker tries to maximize the utility, the SOM replaces *informativity* with the notion of *Expected Success* (given by the expected value of the probability of L successfully guessing the actual height h_0), and replaces the cost function with its marginalization over all possible heights $h > \theta$. Also note that the cost function is defined slightly differently from the one in (6). Here c_{som} is still a parameter attributed to the use of u_1, but the probability of the speaker uttering u_1 is now determined by the choice of θ. Therefore the cost function defined in (7) should be more appropriately interpreted as an *expected* cost function.

(7)   With the cost parameter c_{som},

$$U_{\text{som}}(\theta) = ES(\theta) - \text{Cost}(u)$$

$$= \int_{-\infty}^{\theta} \phi(h)\phi(h|u_0,\theta)dh + \int_{\theta}^{\infty} \phi(h)\phi(h|u,\theta)dh - \int_{\theta}^{\infty} \phi(h) \cdot c_{\text{som}}dh$$

Then following the same spirit as in RSA of modeling real-life agent, the threshold θ is chosen sub-optimally as in (5):

(8) $Pr(\theta) \propto exp(\lambda \cdot U(\theta))$

Finally, according to Degree semantics, the probability of using the utterance u can be given as the probability of $\theta \leq h_0$:

(9) $\sigma(u|h_0) = P(\theta \leq h_0) = \int_{-\infty}^{h_0} Pr(\theta)d\theta$

1.3   A Pragmatic Story for Intensifiers: Bennett and Goodman (2018)

Based on the RSA model, Bennett and Goodman (2018) proposed a purely pragmatic account for degree modifiers. Though agreeing with Klein on that modified adjective phrases have the same meaning as unmodified ones except for a threshold shift, they claimed that intensifiers such as 'very' or 'extremely' give rise to the shift in a non-compositional way, by simply changing the cost function Cost(u). It then follows that intensified adjectives such as 'very tall' and 'extremely tall' have the same semantic meaning as the vanilla form 'tall', except for (different amount of) shifts in the thresholds induced by two intensifiers ('very' and 'extremely') with different costs. And since 'extremely' is more costly than 'very', the threshold for 'extremely tall' is higher than that of 'very tall'. The cost parameters $c_{\text{rsa}}/c_{\text{som}}$ may depend on the length of the intensifier (longer words cost more than shorter ones) and the frequency (rarer words are harder to access, hence also cost more), etc. This account can be easily implemented using the RSA model, and Bennett & Goodman have already shown its robustness against experimental data. In this paper, we want to further compare their account with Klein (without any preference a priori), by implementing both of them in both RSA and SOM. We also hope to gain more insights for the conceptual and mathematical constructions of the two Bayesian models during the process.

1.4   *Pros* and *Cons* for Intensifier Accounts

Before proceeding to the project, we would conclude the introduction section with some theoretical arguments for or against each of the two accounts of degree modifiers. The key debate, as indicated above, lies on whether the intensifiers contribute to the threshold-shift compositionally, with non-vacuous lexical semantics. As Bennett & Goodman pointed out, though it is intuitive to encode the strengths of intensification into the lexical meanings of degree modifiers, it faces certain obstacles. First and foremost, there is a large multitude of degree modifiers with great potential for language production. For example, adverbs like *ridiculously* normally do not indicate an intensifying reading, but when used in *ridiculously tall*, we can easily construe it as an intensification of *tall*. In this sense, to provide lexical semantics for each intensifier would greatly affect theoretical parsimony.

On the other hand, Bennett & Goodman's account also suffers from certain theoretical deficiencies. For one thing, it cannot exclude the possibility that the cost induced by an intensifier has something to do with its lexical meaning. As is mentioned in Sect. 1.2, the cost of an intensifier may depend on its length and frequency, but it is reasonable to argue that the word is rarely used because of its relatively extreme meaning. Hence before we accept this account, we need to gain more evidence regarding the causal direction. Moreover, since the account is purely pragmatic and non-compositional, it faces direct objections with respect to compositionality. Consider the following sentences:

(10)   a. Ronald is not *extremely tall*.

    b. Ronald is *extremely tall* and *quite smart*.

Intuitively, (10a) means Ronald's height doesn't exceed the average height saliently, but may indicate that he can be relatively tall (or serves as a euphemism to say he's not tall). However, with Bennett & Goodman's account, the intensifier *extremely* doesn't contribute to the semantic meaning at all. But since it significantly strengthened the meaning of the adjectival phrase *not tall* (with a high cost), we can derive the meaning that 'Ronald is extremely not tall', which contradicts the general intuition that *extremely* is interpreted in the scope of negation. Also, (10b) means that Ronald's height saliently exceed the average, and his intelligence is somewhat above average. However, if we construe the intensification as purely pragmatic, we lose the binding between *extremely* and *tall* as well as *quite* and *smart*, and fail to derive the correct reading. Last but not least, notice that intensified forms of absolute adjectives such as '*very late*' usually have a relative reading, as opposed to their plain forms with absolute ones. This transition is not likely to be captured by Bennett & Goodman's account, following which the purely pragmatic intensifier 'very' would not introduce a qualitative difference.

Again, this paper doesn't have preference for either account. In contrast, with a (roughly) semantic vs. pragmatic contrast between the two accounts, it is a tempting idea to reconcile the two accounts in one quantitative model to better capture the behavior of gradable adjectives. As mentioned previously, the project incorporates both accounts into both SOM and RSA, presenting a four-way comparison between theories and models, with a baseline provided by experimental data. Unfortunately, neither RSA nor SOM are in position to make predictions for absolute adjectives at this moment, as their reinterpretations turn out to be diverging. Therefore, in the remaining part of the paper we will only focus on relative adjectives and leave absolute cases for future research (for more discussions, see also Sect. 2.5). The project will be introduced in detail in Sect. 2, with further discussions and future directions in Sect. 3.

2   Project: Data and Results

2.1   Goal

The goal of this project is to test two different accounts proposed by Klein and Bennett & Goodman, respectively, for the interpretation of degree modifiers, embedded in both SOM and RSA. We expect it to provide empirical evidence for/against either of the interpretations, and to provide insights for the comparison between the two Bayesian models. Specifically, we want to see how well the quantitative predictions from different models following different interpretations fit with empirical data obtained from the experiment conducted by Leffel et al. (2018).

2.2   Model and Data

Leffel et al. (2018) measured participants' agreement with sentences such as "Ronald is tall" given Ronald's exact height. They tested both 'tall' and 'very tall' (among other constructions), for 13 different heights from 5 ft 3 in. (160 cm) to 6 ft 10 in. (208 cm). In a unit setting, participants are given a combination of some facts (e.g. 'Ronald is 70-in. tall.') and an utterance (e.g. 'Ronald is very tall.'), and are required to adjust a slider to indicate their degree of agreement in a scale of 0 to 100. For our purpose, we interpret these judgments as reflecting the probability that a sentence is true, i.e. the probability that $\theta \leq h$.[2] For SOM, this translates naturally as the cumulative distribution function of θ, $\int_{-\infty}^{h} Pr(\theta)d\theta$. For RSA, we will translate this as the posterior cumulative distribution function of θ, as inferred by the pragmatic listener: $\int_{-\infty}^{h} \rho(\theta|u_1)d\theta$.

We tested all combinations of the two theoretical claims (Klein vs. Bennett & Goodman) and two probabilistic models (SOM vs. RSA), against the median judgment for each point of the scale and each construction. While in principle the models make predictions about individual speakers rather than a population (particularly the RSA), we chose to model the latter for simplicity and because the individual data was rather noisy. The median was preferred to the mean, as it is more robust against outliers and because the mean would not converge to 0% or 100% for extreme values[3]. In each case, we started by adjusting the model parameters (cost c for 'tall', λ, prior on heights) to fit the data for 'tall', and then evaluated the *best* possible fit for '*very tall*'. All results are presented in Fig. 1.

---

[2] Leffel et al. (2018) interpret their results as reflecting not just truth but also pragmatic felicity (i.e. truth of the sentence together with its implicatures). However the implicatures they discuss only surfaced for more complex sentences (involving negation), and should therefore do not affect our interpretation of the simpler sentences discussed here.

[3] Here we leave a more in-depth evaluation of the model for later. Ideally, we would like to fit not just the median but the data from each participant (as both SOM and RSA are originally individual-level models), and in a more systematic way.

(i) **Klein + SOM**

To fit the data for '*tall*', we chose the prior distribution $\phi(h)$ as a normal distribution with parameters $\mu = 68.5$ in., $\sigma = 3.7$ in. The degree of rationality was $\lambda = 1.2$, and the cost parameter for *tall* was $c_{\mathrm{som}} = 0.2$.

According to Klein's account for degree modifiers, intensified adjectives such as *very tall* are interpreted as *tall compared to the set of tall people*. It can be incorporated into SOM by using the posterior distribution on heights after an utterance of *tall* as the prior distribution $\phi'(h)$ for the heights of *tall people*:

(11) $\phi'(h) = \phi(h|\theta, u_1) \propto \phi(h) \int_{-\infty}^{h} \frac{Pr(\theta)}{1-\Phi(\theta)} d\theta$

To simplify the further computation, we approximated this distribution with a Gaussian in the next steps. Then combining (7), (8), (9) and (11), we can derive the distribution of the threshold θ' for *very tall*, which can then be integrated to derive a model of participants' median judgments as the probability $P(\theta' < h)$. The cost parameter for this second iteration of the algorithm was c'_{som}, and was meant to reflect the cost of adding *very* to the sentence. With constraint $c'_{\mathrm{som}} \geq 0$, the optimal choice ended up being 0 (see Fig. 1a for details).

(ii) **Bennett & Goodman + SOM**

According to Bennett & Goodman, intensifying degree adverbs shift the threshold just because they increase the cost of utterances. Therefore, we could translate this account into the language of SOM simply by increasing the value of the cost parameter to $c_{\mathrm{som}} + c'_{\mathrm{som}}$, where c'_{som} is the additional cost caused by *very*. Here, as shown in Fig. 1b, the optimal choice was $c'_{\mathrm{som}} = 1.8$. Other parameters were identical to what we used to implement Klein's account.

(iii) **Klein + RSA**

To fit the data for *tall* with the RSA, we chose the prior distribution $\phi(h)$ again as a normal distribution with parameters $\mu = 69$ in., $\sigma = 3.7$ in., the degree of rationality was $\lambda = 4.8$, and the cost for *tall* was $c_{\mathrm{rsa}} = 0.85$.

The RSA naturally provides the posterior distribution $\rho(h|u_1)$ for the heights of *tall people*:

(12) $\phi'(h) = \rho(h|u_1) \propto \phi(h) \int_{-\infty}^{h} \frac{Pr(\theta)}{1+e^{\lambda c_{\mathrm{rsa}}} \cdot (\int_{\theta}^{\infty} \phi(h')dh')^{\lambda}} d\theta$

Then combining (12) with (4), (5), (6), we derive Klein's predictions for *very tall* within the RSA model. A cost $c'_{\mathrm{rsa}} = 1/3 c_{\mathrm{rsa}} = 0.28$ gave close to optimal results.

(iv) **Bennett & Goodman + RSA**

As in (ii), Bennett & Goodman's account of *very* only requires increasing the cost to $c_{\mathrm{rsa}} + c'_{\mathrm{rsa}}$. Setting all parameters as in (iii) gave good results.

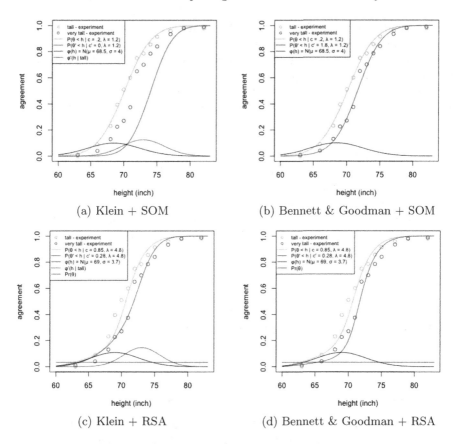

Fig. 1. Best fit for each model, compared to the median response in Leffel et al. (2018)

2.3 Results

Figure 1 shows the results of comparing experimental data with model predictions described above. Figure 1a indicates that even with a minimal cost $c'_{\text{som}} = 0$ for the intensifier, the prediction of Klein's account implemented within SOM shifts the threshold of '*very tall*' far to the right. All the other three combinations gave good approximations of the data, although Bennett & Goodman's account within SOM required a very high cost $c'_{\text{som}} = 1.8$ compared to $c_{\text{som}} = 0.2$ for the full unmodified sentence. Both accounts of '*very*' could be implemented in RSA with a reasonable cost for *very*.

2.4 Discussion

In Fig. 1, we chose to present the radical case $c'_{\text{som}} = 0$ because any larger value of c'_{som} shifts the curve further right (as higher cost will generally give rise to a higher prediction of the optimal threshold); and since this radical case has already yielded a result far to the right, it indicates that SOM + Klein cannot make satisfying predictions for *very tall*. Second, for the combination of SOM + RSA as shown in Fig. 1b, though it exhibits an accurate prediction, it requires an additional cost parameter $c'_{\text{som}} = 1.8$. If we follow the assumption that cost is proportional to the length of expression as measured by the number of words (see Lassiter and Goodman 2014), then compared to the cost $c_{\text{som}} = 0.2$ set to fit the data of *tall*, which is induced by the sentence *Ronald is tall*, c'_{som} is impractically large. Therefore we conclude here that SOM is relatively unfit for the interpretations of intensified (relative) adjectives. The reasons we suspect here are that (1) The major predictions made by SOM concentrate on the speaker's level S (one level below RSA), resulting in its 'over-reaction' to the additional iteration for the implementation of Klein's account, and (2) SOM's low sensitivity to the cost parameter (presented as an advantage of this model in Qing and Franke 2014) requires a higher value of the additional cost in order to accommodate Bennett & Goodman's account. On the other hand, Fig. 1c and d show that RSA does give good predictions with both accounts, and with relatively practical cost parameters $c_{\text{rsa}} = 0.85$ and $c'_{\text{rsa}} \approx \frac{1}{3}c_{\text{rsa}}$. Note that our results offer an independent assessment of Bennett and Goodman (2018), as they only tested their models against a point-wise estimate of the posterior (but for multiple intensifiers), while we tested the model's ability to fit the full posterior distribution (though only for *very*). Interestingly however, the present results do not clearly advocate between Klein and Bennett & Goodman.

2.5   Some Remarks on Absolute Adjectives

As briefly mentioned in Sect. 1.4, the implementations of RSA and SOM indicate that neither of them are in the position of providing predictions for absolute adjectives. It is especially disappointing since in principle, absolute adjectives like '*late*' would help further distinguish between the proposals of Klein and Bennett & Goodman. The problems, as we will see, come from the tension between the computational model and experimental data. One particularly difficult aspect of the data is that participants do not treat the absolute adjective '*late*' as purely minimum-standard, though they still place a significant probability mass on the bottom of the scale. The distribution of θ is a mixed distribution with roughly 2/3 of the probability mass on the bottom point and the remaining 1/3 distributed continuously to the right of this point. This is in line with the more classical accounts of Burnett (2014) who argue that absolute adjectives show some tolerance, but only in one direction. But this is precisely the obstacle to a satisfying computational model. We observe that both Lassiter and Goodman (2014) and Qing and Franke (2014) take β-functions as the priors for absolute adjectives, as they effectively describe the *closure* feature of degree scales with a significant probability mass at the lower- or upper- bound. However, in order to reinterpret them for the test against experimental data, we need to take the cumulative distribution function such as (6) and (9). This operation results in predictions for absolute adjectives that are either a continuous distribution without any qualitative difference with relative adjectives, or a diverging, all-or-nothing step function. Specifically, the RSA cannot derive the discrete probability mass at the closure point as it only derives continuous distribution, while the SOM either derives a purely minimum-standard or a purely relative reading (depending on how much probability mass of the prior is located at the bottom of the scale).

In order to overcome this challenge, we believe a re-examination of the conceptual and mathematical constructions of the current models is needed. Further, we suspect that the problem can at least be mitigated once we take a closer look at the structures of their prior distributions. We think the difference in interpretations of a relative adjective and an absolute adjective may originate from the characteristics of an agent's prior knowledge associated with the them, as the former can be intuitively described as a normal distribution, the latter indicates a natural discontinuity at the lower point (in the case for '*late*', say, the prescribed arrival time for a meeting). We leave a detailed investigation for future works, and we hope it can lead to other revelations of the nature of vagueness in language.

3   Conclusion

We presented a four-way comparison between (i) two theories of degree modifiers, i.e. Klein (1980) and Bennett and Goodman (2018), and (ii) two recent probabilistic models for scalar adjectives, i.e. RSA (Lassiter and Goodman 2014) and SOM (Qing and Franke 2014), in dealing with degree modifiers. The comparison was conducted by testing the 2×2 combinations of theories and models on the experimental data from Leffel et al. (2018). The results showed us that SOM, due to its 'over-reaction' to a second derivation for the intensifier, and its low sensitivity to the cost parameter, does not make a good prediction for the meaning of intensifiers; whereas RSA presented us with relatively good results. Meanwhile, the results did not show preference between Klein and Bennet & Goodman. In particular, we could imagine implementing Bennett & Goodman's flexible account of various intensifiers by varying the cost in the second derivation involved in Klein's account (so '*extremely tall*' would also mean 'tall among tall people', but with a higher cost than '*very tall*'). In fact, this is virtually the strategy we used trying to find the best fit for Klein's account with both RSA and SOM models.

In this paper, we only discussed the relative standard adjective '*tall*'. Turning to minimum-standard absolute adjectives such as '*late*' would in principle help further distinguish between the proposals of Klein and Bennett & Goodman. Crucially, the correct account should explain for the fact that '*late*' is minimum-standard while '*very late*' is relative. Nevertheless, neither the SOM nor the RSA was in position to fit the data for *late* presented in Leffel et al. (2018), forcing us to leave this question for future research. Also, besides intensified constructions, it would be interesting to see if there is a model that can cover other complex constructions such as negated forms ('*not tall*', '*not very tall*') and comparative forms ('*taller*', '*later*'), which typically manifest the interaction between implicatures and the absolute/relative distinction as discussed in Leffel et al. (2018).

Acknowledgements. The author would like to thank Dr. Alexandre Cremers for supervising the project. I'm also grateful to the anonymous reviewers for their valuable feedback. Last but not least, many thanks to ESSLLI 2018 Grant Committee and EACL for the student grants, and Springer for sponsorship and prizes.

References

Bennett, E.D., Goodman, N.D.: Extremely costly intensifiers are stronger than quite costly ones. Cognition **178**, 147–161 (2018)

Burnett, H.: A delineation solution to the puzzles of absolute adjectives. Linguist. Philos. **37**(1), 1–39 (2014)

Egré, P., Bonnay, D.: Vagueness, uncertainty and degrees of clarity. Synthese **174**(1), 47–78 (2010)

Franke, M., Jäger, G.: Probabilistic pragmatics, or why Bayes' rule is probably important for pragmatics. Zeitschrift für sprachwissenschaft **35**(1), 3–44 (2016)

Kennedy, C.: Vagueness and grammar: the semantics of relative and absolute gradable adjectives. Linguist. Philos. **30**(1), 1–45 (2007)

Kennedy, C., McNally, L.: Scale structure, degree modification, and the semantics of gradable predicates. Language **81**, 345–381 (2005)

Klein, E.: A semantics for positive and comparative adjectives. Linguist. Philos. **4**(1), 1–45 (1980)

Lassiter, D., Goodman, N.D.: Context, scale structure, and statistics in the interpretation of positive-form adjectives. In: Semantics and Linguistic Theory, pp, 587–610 (2014)

Leffel, T., Cremers, A., Gotzner, N., Romoli, J.: Vagueness in implicature: the case of modified adjectives. Under Rev. Semant. **36**, 317–348 (2018)

Qing, C., Michae, F.: Gradable adjectives, vagueness, and optimal language use: a speaker-oriented model. Semant. Linguist. Theor. **24**, 23–41 (2014)

Smith, N.J.J.: Vagueness and Degrees of Truth. Oxford University Press, Oxford (2008)

Williamson, T.: Vagueness. Routledge, London (2002)

Author Index

Printed in the United States
By Bookmasters